South Africa's Emergent Middle Class

This book is drawn from diverse studies that grapple with black middle class experiences in contemporary and historical South Africa. The chapters present research from diverse disciplines, and tackle issues related to being black and middle class, using both quantitative and qualitative approaches. Like many other social phenomena, the black middle class concept is seen as complex and not easy to pin down. As a result, conceptualisations from these chapters are dynamic and relevant for understanding the position of the black middle class in contemporary South African society. An interesting dynamic explored by contributors is the critical engagement with the usually reductionist notions of black middle class experiences as ahistorical, homogenous experiences of a group of conspicuous consumers. These limiting notions are unpacked and repositioned in how the book is structured.

This book was published as a special issue of *Development Southern Africa*.

Grace Khunou is Associate Professor in the Sociology Department at the University of Johannesburg, South Africa. She writes creatively and academically, and has published articles in peer-reviewed journals and written chapters in edited books. She has also presented papers at international and local conferences. Her research interests are focused on gender and social policy, the black middle class, narrative research, and transformation in Higher Education. She is a member and convener of the health working group for the South African Sociological Association, and a member of the International Sociological Association. She has recently co-edited a special issue on Father Connections for the *Open Family Studies Journal*, and a special issue on the emergent black middle class for the journal *Development Southern Africa*.

South Africa's Emergent Middle Class

Edited by
Grace Khunou

Routledge
Taylor & Francis Group

LONDON AND NEW YORK

First published 2016
by Routledge

2 Park Square, Milton Park, Abingdon, Oxfordshire OX14 4RN
711 Third Avenue, New York, NY 10017

Routledge is an imprint of the Taylor & Francis Group, an informa business

First issued in paperback 2017

British Library Cataloguing in Publication Data
A catalogue record for this book is available from the British Library

ISBN 13: 978-1-138-96044-2 (hbk)
ISBN 13: 978-1-138-30636-3 (pbk)

Typeset in Times
by RefineCatch Limited, Bungay, Suffolk

Publisher's Note
The publisher accepts responsibility for any inconsistencies that may have
arisen during the conversion of this book from journal articles to book chapters,
namely the possible inclusion of journal terminology.

Disclaimer
Every effort has been made to contact copyright holders for their permission to
reprint material in this book. The publishers would be grateful to hear from any
copyright holder who is not here acknowledged and will undertake to rectify
any errors or omissions in future editions of this book.

Contents

Citation Information

The chapters in this book were originally published in *Development Southern Africa*, volume 32, issue 1 (January 2015). When citing this material, please use the original page numbering for each article, as follows:

Editorial

South Africa's emergent middle class
Grace Khunou
Development Southern Africa, volume 32, issue 1 (January 2015) pp. 1–2

Chapter 1

Growth of the middle class: Two perspectives that matter for policy
Justin Visagie
Development Southern Africa, volume 32, issue 1 (January 2015) pp. 3–24

Chapter 2

The emergent middle class in contemporary South Africa: Examining and comparing rival approaches
Ronelle Burger, Cindy Lee Steenekamp, Servaas van der Berg & Asmus Zoch
Development Southern Africa, volume 32, issue 1 (January 2015) pp. 25–40

Chapter 3

Understanding consumption patterns of the established and emerging South African black middle class
Ronelle Burger, Megan Louw, Brigitte Barbara Isabel de Oliveira Pegado & Servaas van der Berg
Development Southern Africa, volume 32, issue 1 (January 2015) pp. 41–56

Chapter 4

Life chances and class: Estimating inequality of opportunity for children and adolescents in South Africa
Asmus Zoch
Development Southern Africa, volume 32, issue 1 (January 2015) pp. 57–75

Chapter 5

Rethinking Bundy: Land and the black middle class – accumulation beyond the peasantry
Nkululeko Mabandla
Development Southern Africa, volume 32, issue 1 (January 2015) pp. 76–89

Chapter 6

What middle class? The shifting and dynamic nature of class position
Grace Khunou
Development Southern Africa, volume 32, issue 1 (January 2015) pp. 90–103

Chapter 7

'Growing up' and 'moving up': Metaphors that legitimise upward social mobility in Soweto
Detlev Krige
Development Southern Africa, volume 32, issue 1 (January 2015) pp. 104–117

Chapter 8

Food, malls and the politics of consumption: South Africa's new middle class
Sophie Chevalier
Development Southern Africa, volume 32, issue 1 (January 2015) pp. 118–129

For any permission-related enquiries please visit:
http://www.tandfonline.com/page/help/permissions

Notes on Contributors

Ronelle Burger is an Associate Professor in the Department of Economics at Stellenbosch University, South Africa. She has a PhD in Economics from the University of Nottingham, UK. She co-authored *Enabling Financial Sector Development in African Countries: Case Studies in Four SADC Countries* (2006), and has contributed chapters to several edited collections on post-apartheid South Africa.

Sophie Chevalier is Associate Professor in the Department of Sociology and Anthropology at the Université de Franche-Comté, Besançon, France. Her research interests are centred on the political anthropology of economics, including consumption practices, articulation in both the private and public spheres, and social class and race. Her research is specifically focused on urban environments in Europe and South Africa.

Brigitte Barbara Isabel de Oliveira Pegado is a Researcher in the Department of Economics at Stellenbosch University, South Africa. Her research is focused on econometrics, development economics, and behavioural economics.

Grace Khunou is Associate Professor in the Sociology Department at the University of Johannesburg, South Africa. She writes creatively and academically, and has published articles in peer-reviewed journals and written chapters in edited books. She has also presented papers at international and local conferences. Her research interests are focused on gender and social policy, the black middle class, narrative research, and transformation in Higher Education. She is a member and convener of the health working group for the South African Sociological Association, and a member of the International Sociological Association. She has recently co-edited a special issue on Father Connections for the *Open Family Studies Journal*, and a special issue on the emergent black middle class for the journal *Development Southern Africa*.

Detlev Krige is Senior Lecturer in Anthropology at the University of Pretoria, South Africa. His research is focused on financialisation, gift giving, and economic anthropology.

Megan Louw is a Researcher in the Department of Economics at Stellenbosch University, South Africa. Her research is centred on development and welfare economics.

Nkululeko Mabandla is a Research Officer in the Centre for African Studies at the University of Cape Town, South Africa, and a Doctoral Fellow at the Institute for Humanities in Africa. His research interests include the fields of political economy, post-colonial theory, language, and history. His current research is on Chinese traders in South Africa's rural towns.

NOTES ON CONTRIBUTORS

Cindy Lee Steenekamp is a Researcher in the Centre for International and Comparative Politics, and Director of the Graduate School of Arts and Social Sciences, at Stellenbosch University, South Africa. Her research interests include political behaviour and culture, social capital, and survey research.

Servaas van der Berg is Professor and NRF Chair in Social Policy in the Department of Economics at Stellenbosch University, South Africa. His research interests include economic development, poverty, income distribution, social policy, and education. He consults for government departments and international organisations such as the World Bank, UNDP, Unicef, and SACMEQ/UNESCO.

Justin Visagie is Director of Economic Planning and Research for the Department of Economic Development, Environmental Affairs and Tourism in the Eastern Cape region of South Africa. He has a PhD in Economics from the University of Kwa-Zulu Natal, Durban, South Africa.

Asmus Zoch is a PhD student in the Department of Economics at Stellenbosch University, South Africa, and a Research Fellow with Policy Design and Evaluation Research in Developing Countries (PODER). He is also associated with Research on Socio-Economic Policy (RESEP), as a Researcher working on measurement error, school to work transition, income mobility, and inequality of opportunity.

Introduction: South Africa's emergent middle class

Given the significance of the black middle class for the consolidation of South Africa's democracy, studies on class have focused on the black middle-class experience. However, as with many other social phenomena, conceptualisations from these studies have differed considerably. The challenge is the tendency of many of these conceptualisations to reduce black middle-class experiences to the ahistorical, homogeneous experiences of a group of conspicuous consumers. This is based on the limiting notion that black experience is traditional and uncomplicated, and this approach is then extrapolated to the experiences of the black middle class.

This issue brings together eight dynamic articles on the black middle class in South Africa. The articles present research from diverse disciplines and tackles issues related to being black and middle class from both quantitative and qualitative stands. The quantitative articles provide a critical reading of conceptualisations of black middle-classness. For example, the article entitled 'Growth of the middle class: Two perspectives that matter for policy' by Visagie compares income changes of the relatively affluent 'middle class' with those of households in the literal middle of the income spectrum. This article shows that in the affluent middle there has been significant racial transformation and growth of the black middle class. However, households in the actual middle of the income spectrum have experienced the lowest income growth of all groups since 1993. He also argues that both perspectives are crucial in the pursuit of a more equitable path of development and have important implications for policy. On a more critical note, the article 'The middle class in contemporary South Africa: Comparing rival approaches' by Burger et al. cautions against overoptimistic predictions based on the growth of the black middle class, suggesting that while the surge in the black middle class is expected to help dismantle the association between race and class in South Africa, critical analysis suggests that notions of identity may adjust more slowly to these new realities and, consequently, racial integration and social cohesion may emerge with a substantial lag. Another article by Burger et al., 'Understanding the consumption patterns of the established and emerging South African black middle class', foregrounds the importance of critical analysis in class studies by moving away from analysis of black middle-classness that leans either on the theory of conspicuous consumption or culture-specific utility functions. This article interestingly argues that households new to the middle class or uncertain of continued class membership are viewed as vulnerable. On the same note, in the article entitled 'Life chances and class: Estimating inequality of opportunity for children and adolescents in South Africa', Zoch provides a multivariate analysis that highlights the importance of class membership for schooling outcomes and labour market prospects of a child.

The qualitative articles, on the other hand, begin to illustrate the dynamics that identity and other social markers play in how black middle-classness is experienced. In a critique of notions that suggests black middle-classness is a post-1994 phenomenon, Mabandla foregrounds the idea of an earlier existence of a black middle class through the article 'Rethinking Bundy: Land and the black middle class – accumulation beyond the peasantry'. Through assessment of historical data on the black middle class in Mthatha, this article illustrates that South Africa's black middle class has considerable time depth that can be traced through the reading of Bundy's 'peasantry'. Khunou's article 'What middle class? The shifting and dynamic nature of class position' takes Mabandla's argument further by looking at how middle-classness for blacks during apartheid was marred with constant shifts related to the socio-economic and political impermanence of class position, and concludes that black middle-classness is complex and heterogeneous and thus cannot be understood without historical analysis. Krige's article entitled '"Growing up" and "moving up": Metaphors that legitimise upward social mobility in Soweto' moves on to provide a thought-provoking analysis of accusations of materialism and conspicuous consumption levelled against the urban black middle class and argues for what is referred to as a renewal of cultural practices in which private wealth can legitimately be converted into social wealth. In the article 'Food, malls and the politics of consumption: South Africa's new middle class', Chevalier examines middle-class interaction in shared social spaces that were previous segregated and argues that South Africans are willing to experiment beyond the boundaries of their native communities and there is an emergent national middle-class culture, but there are marked regional differences and nothing yet that would amount to 'creolisation'.

Grace Khunou
Editor, *Development Southern Africa*

Growth of the middle class: Two perspectives that matter for policy

Justin Visagie

Two very different pictures emerge when one compares income changes of the relatively affluent 'middle class' with those of households in the literal middle of the income spectrum. In the affluent middle there has been significant racial transformation and growth of the 'black middle class'. However, households in the actual middle of the income spectrum have experienced the lowest income growth of all groups since 1993. Both perspectives are crucial in the pursuit of a more equitable path of development and have important implications for policy design.

1. Introduction

Any discussion of changes in the income status of the middle class in South Africa depends on how the term 'middle class' is defined. In previous work (Visagie & Posel, 2013), I have shown how the common practice of defining the middle class as 'the relatively affluent' is inconsistent with a focus on those in the actual middle of the distribution of income. Widespread poverty and income inequality in South Africa imply that households which are in the middle in terms of their income status are still a long way off from what is deemed a middle-class standard of living. In fact, those in the middle still are quite poor.

However, holding these two perspectives (the median versus affluent middle class) in tension provides novel insight into the nature of South African income growth over the first 15 years of democracy, with important implications for policy. In this article, I track changes that have taken place in these two different 'middle class' groups since 1993 and ask: have they become better off? The answer to this question provides novel insight into changes in the income landscape in the country.

The remainder of the article is structured as follows. Section 2 provides a brief context of the importance of the middle class and of relevant policy in South Africa. Section 3 discusses an 'affluence-based' and a 'median-based' definition of the middle class and outlines the household survey income data used in the study. Section 4 presents descriptive findings on changes in the size and demographic composition of the affluent middle class over the period 1993–2008, whereas Section 5 contrasts these findings with an evaluation of income growth of households in the actual middle. The final section draws out the implications for policy design.

2. The importance and the politics of the middle class

The middle class is considered to provide a range of key functions for the economic growth and development of a country. These functions include: providing a base of growing human capital (Galor & Zeira, 1993; Perotti, 1996; Sokoloff & Engerman, 2000), through attitudes and behaviours specific to the middle class (such as savings and entrepreneurship) (Landes, 1998; Doepke & Zilibotti, 2005), as the primary driver of domestic consumption (Brown, 2004; Bhalla, 2007; Pressman, 2007; Kharas & Gertz, 2010), holding government accountable (Kharas & Gertz, 2010; African Development Bank, 2011) and promoting political stability (Barro, 1999; Easterly, 2001, 2007).

However, the above summary of the importance of the middle class for economic development hides fairly obvious contradictions in the way in which this group is used and defined across each of these studies. Such ambiguity needs to be more carefully considered and acknowledged, particularly in light of the implications for income inequality or inclusive growth in developing economies (Visagie & Posel, 2013).

The idea of the middle class has its genesis in the sociology literature, particularly the seminal works of Weber (1961) and Marx (1974). From a sociological perspective, the middle class is defined by their economic function. In practice, however, class has been measured using a wide variety of indices (and in various combinations), such as measures of income, household assets, education levels and occupation (Pressman, 2007; Birdsall, 2010; Atkinson & Brandolini, 2013). The economics literature makes widespread use of income (or expenditure) to very pragmatically measure and define the middle class (Ravallion, 2010; Atkinson & Brandolini, 2013). I follow on from previous work in Visagie & Posel (2013) in exploring two common measures of the middle class from the economics literature: those in the actual middle of the income distribution (median-based approach), and those with a middle-class standard of living (affluence-based approach). However, widespread poverty and very high income inequality in South Africa imply that these two 'middle class' groups do not overlap. The middle class in South Africa cannot be defined as the relatively affluent and simultaneously as the middle majority (Visagie & Posel, 2013).[1]

This tension equally applies to thinking about the politics of the middle class in South Africa. During apartheid, class cleavages were legislated according to race with middle-class affluence promoted amongst whites (Seekings & Nattrass, 2005). Individuals of 'colour' were denied business opportunities, dispossessed of property, forcibly relocated and barred from occupational advancement. The rise of the African National Congress to power in 1994 marked the end of such a coerced relationship between race and class. Nevertheless, after 46 years of apartheid, the relative economic advantage of whites within South Africa had been firmly established. In 1993 the median per-capita income for whites was more than nine times larger than that for Africans (author's own estimates; Project for Statistics on Living Standards and Development [PSLSD] 1993). A central policy goal within the democratised administration is therefore a rise in the economic status amongst previously disadvantaged groups, and particularly amongst Africans. This is seen in Affirmative

---

[1]This contradiction between the middle majority and middle-class affluent is not unique to South Africa. Ravallion (2010:445) discusses a wide variety of measures of the middle class across a number of developing countries and concludes that 'Western notions of the middle class are of little obvious relevance to developing countries'.

Action and Black Economic Empowerment (BEE) legislation. However, a common criticism of BEE policy is that it fosters the growth of a black elite whilst failing to empower the black majority (Ponte et al., 2007; Southall, 2007, Hoffman, 2008). In other words, while BEE policy targeted the creation of a 'black middle [and upper] class', such policy does little for the average black household in the actual middle.

The cause of the middle majority, however, does find traction in South African macroeconomic strategy, which consistently makes reference to the importance of reducing income inequalities. This is emphasised in the Reconstruction and Development Programme, which places 'the problems of poverty and gross inequality' as foremost challenges to South Africa's development (RDP, 1994:4). The Growth, Employment and Redistribution strategy that followed on from the RDP again recognises the need to address inequality – although lowering inequality is limited to an outcome of economic growth. Replacing the Growth, Employment and Redistribution strategy, the Accelerated and Shared Growth Initiative for South Africa re-emphasises that growth itself is unsustainable without a reduction in historical inequalities; that is, reducing inequality can be an outcome and a means to achieve economic growth. Similarly, the New Growth Path and the National Development Plan explicitly acknowledge inequality as a core development challenge and intentionally plan for a more 'inclusive' society. The point here is that the growth and development of the middle majority is implicit within South African macroeconomic policy, albeit outside the narrow policy reach of BEE (which applies narrowly only to the most affluent).

3. Definitions and data

In this study I track changes that have taken place in both the median-based and affluence-based middle-class groups. Both groups of 'middle class' households are of analytical interest and become even more useful when contrasted against each other. Hence two competing perspectives of the middle class are investigated in this study.[2]

In analysing the actual middle of the income distribution in South Africa, I define the middle strata as households (and individuals) which fall within an interval of 50 to 150% of the median per-capita household income. This is a common approach within the international literature, particularly in developed economies (see Thurow, 1987; Davis & Huston, 1992; Pressman, 2007). Using data from the 2008 National Income Dynamics Study (NIDS), the middle strata are identified by a per-capita household income of between R380 and R1140 per capita per month in 2008 prices.

To identify individuals with a middle-class lifestyle or the 'affluent middle class', I use an income threshold of between R1400 and R10 000 per person per month (after-tax earnings in 2008 prices). This corresponds to the average earnings for households in which the highest income earner is in a typically 'middle-class' occupation in the NIDS 2008.[3] The upper bound of R10 000 places only the wealthiest 2% of individuals from the South African income distribution into the upper class (based on NIDS 2008 data). The lower bound of R1400 also roughly corresponds to a

---

[2]Refer to Visagie & Posel (2013) for an extended discussion of each definition and how these definitions were derived.
[3]Middle-class occupations are derived from the International Standard Classification of Occupations and are defined as 'legislators, senior officials and managers', 'professionals, associate professionals and technicians' and lastly 'clerks'.

commonly utilised lower income threshold in the international literature of $10 per day (Bhalla, 2007; Birdsall, 2010; Kharas & Gertz, 2010). Those in the lower class (individuals with less than R1400 per capita per month) are further divided into the poor (<R515) and non-poor (R515 to R1399) using a basic cost-of-needs poverty line (see Hoogeveen & Özler, 2006; Posel & Rogan, 2009; Leibbrandt et al., 2010).

The data come from three nationally representative cross-sectional datasets for South Africa. These are the 1993 PSLSD; the 2000 Income and Expenditure Survey (IES), which can be combined with the 2000 Labour Force Survey (LFS) as they surveyed the same group of households; and lastly the NIDS 2008. The NIDS provides a later wave of data for 2010/11, but there is a high rate of sample attrition for individuals at the top of the income distribution. Despite reweighting the data to correct for this attrition, Finn et al. (2012) find that there remains a downward bias in incomes at the top, and hence wave 2 data from the NIDS has not been included. Nevertheless, my own preliminary examination of the NIDS 2010/11 data suggests little difference compared with estimates of the middle class in 2008 (discounting the possible effects of sample attrition).

These particular cross-sectional datasets have been chosen because they each collect detailed information on household income (and expenditure), as opposed to merely a one-shot question on total household income (as has been used previously in studies of the middle class; Schlemmer, 2005; Van der Berg, 2010) as well as providing detailed labour market and demographic information. Total household income is computed from the collation of individual incomes from a comprehensive list of sources: namely, income from the labour market, government grant income, income from investments and remittance income. This is a noteworthy improvement on previous estimates of the middle class in South Africa.

Using three different cross-sectional datasets will evidently introduce unwanted complication in the comparison of data across time. These concerns are well discussed in the wider literature on measuring changes in poverty and inequality in South Africa (Bhorat & van der Westhuizen, 2008; Posel & Rogan, 2009; Leibbrandt et al., 2010). There may be some unavoidable bias in comparisons of income data across time. However, three notable features help build some measure of confidence in the results to follow. Firstly, a decline in the poverty headcount ratio between 1993 and 2008 as reported in Table 1 (described by the size of the poor lower class) is corroborated by other studies of income poverty in South Africa – such as the use of AMPS data scaled up to the national accounts aggregates by Van der Berg et al. (2008) or the use of October Household Surveys and General Household Surveys by Posel & Rogan (2009). Secondly, trends in class size and composition are consistent across all three periods and are not merely the construction of a faulty base estimate. Thirdly, the income data are the aggregation of detailed information on the sources of household income in all three datasets.

4. Perspective 1: The affluent middle and the black middle class

4.1 The size of the affluent middle class

Important in the first perspective of affluence-based middle-class development are changes to the size and racial composition of the affluent middle class (R1400 to R10 000 per capita per month in after-tax income from all sources).

6

Table 1: Class size and income share, 1993–2008

	Lower class (<R515)			Lower class (R515 to R1 399)			Middle class (R1 400 to R10 000)			Upper class (>R10 000)			Total population		
	1993	2000	2008	1993	2000	2008	1993	2000	2008	1993	2000	2008	1993	2000	2008
Class size															
Count (millions)	22.7	24.6	25.1	9.1	9.8	11.8	7.7	8.7	10.4	0.4	0.8	1.3	39.9	44.0	48.7
	(0.1)	(0.1)	(0.2)	(0.1)	(0.1)	(0.2)	(0.1)	(0.1)	(0.3)	(0.0)	(0.0)	(0.1)	(0.0)	(0.1)	(0.4)
Percentage share	56.9	56.0	51.7	22.7	22.4	24.2	19.3	19.8	21.3	1.1	1.9	2.8	100	100	100
	(0.3)	(0.2)	(0.5)	(0.2)	(0.2)	(0.4)	(0.2)	(0.2)	(0.5)	(0.1)	(0.1)	(0.3)			
Total income from all sources															
Count (billions)	R5.0	R5.9	R6.2	R7.7	R8.3	R10.0	R26.0	R31.6	R36.6	R7.9	R14.4	R24.5	R46.5	R60.2	R70.6
	(0.0)	(0.0)	(0.1)	(0.1)	(0.1)	(0.2)	(0.4)	(0.4)	(1.1)	(0.6)	(0.7)	(2.4)	(0.7)	(0.8)	(2.4)
Percentage share	10.7	9.8	8.0	16.5	13.7	12.9	55.9	52.5	47.3	16.9	23.9	31.7	100	100	100

Source: PSLSD 1993, IES/LFS 2000, NIDS 2008; author's own estimates.

Notes: Standard errors in parentheses; the data are weighted.

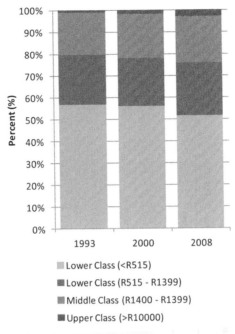

Figure 1: Class population shares, 1993–2008
Source: PSLSD 1993, IES/LFS 2000, NIDS 2008.
Notes: The data are weighted; author's own estimates.

Between 1993 and 2008, an additional 2.7 million individuals were added to the middle class (see Table 1). Hence, in absolute terms there was sizeable growth in number of individuals who attained a middle-class lifestyle. Nevertheless, in relative terms, the size of the middle class increased only slightly ahead of population growth – the population share of the middle class increased from 19.3% of the total population in 1993 to 21.3% in 2008.

This discounts a small share of individuals at the very top of the distribution (i.e. upper class) with incomes in excess of R10 000 per capita per month. Although growing from a low base of less than half a million individuals in 1993, the upper class experienced the largest relative growth of all class groups, trebling in size to 2.8 million individuals over the period (but still accounting for less than 3% of the total population in 2008). Showing strong growth of the upper class (rather than the middle class) illuminates the nature of increases in income inequality experienced in the post-apartheid period (Van der Berg & Louw, 2004; Hoogeveen & Osler, 2006; Leibbrandt et al., 2010).

The vast majority of individuals in South Africa are situated in the lower income classes (poor and non-poor). In 1993, close to 80% of the population fell beneath the middle-class lower bound of R1400 per capita per month (Figure 1). Hence, recognising that the affluent middle approximates the top 20% of individuals in the distribution (when combined with the top-most 3% from the upper class) shows that there is nothing 'middle' about middle-class affluence in South Africa.

The lower class is further divided into the poor and non-poor for further depth of analysis. The proportion of individuals living below the poverty line (R515 per capita

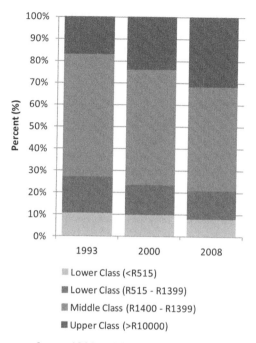

Figure 2: Class income share, 1993–2008
Source: PSLSD 1993, IES/LFS 2000, NIDS 2008.
Notes: The data are weighted; author's own estimates.

per month) declined by 5.2 percentage points between 1993 and 2008. Most of this decrease occurred post 2000, in response to a large expansion in social grant payments. However, in spite of a relative shift away from poverty, the absolute number of individuals living in poverty increased by 2.4 million due to population growth.

The relative size of the middle class is also sometimes measured in terms of the size and share of total income (Pressman, 2007; Palma, 2011). Figure 2 illustrates how the share of total income shifted between income classes between the years 1993, 2000 and 2008. Most striking is a rise in the share of total income accruing to the upper class, at the expense of the remaining income classes. The middle-class income share actually declined from 56% in 1993 to 47% by 2008 – evidence of a 'middle class' income squeeze (see also Palma, 2011).

The fact that the size of the middle class increased only modestly between 1993 and 2008 and that the income share of the middle class actually declined is somewhat surprising in light of other studies of the middle class in South Africa, which generally report strong middle class growth post 1994 (Whiteford & Van Seventer, 2000; Schlemmer, 2005; Van der Berg, 2010). I suggest three potential reasons for this discrepancy (notwithstanding the confounding impact of differences in definition between these studies). The first reason is that there may have been a large absolute increase in the size of the middle class, but the same is not true of a large relative increase in the size of the middle class (or in terms of the income share of the middle class). Secondly, previous studies of the middle class in South Africa have not distinguished between the middle class and the upper class, and the upper class did experience very large growth (albeit from a low base). A third factor is that my estimates of middle class growth are not limited to Africans. In other words, the African middle class may have experienced sizeable growth over the period but the same cannot be said for the middle class as a whole.

4.2. The rise of black affluence

Table 2 describes changes in the racial profile of the affluent middle class in South Africa over the period 1993–2008. An additional 3.1 million Africans were added to the middle class over the period, with a steady increase in size between 1993 and 2000 and between 2000 and 2008. Growth in the number of middle-class Africans is partly driven by growth at the population level – the African population grew by eight million between 1993 and 2008. However, the row percentages in the table for Africans show that growth of the African middle class was in excess of population growth.

In 1993, the middle class consisted of a large white majority Figures 3 and 4. Middle-class whites out-numbered middle-class Africans by almost two to one. However, by 2000 the white middle-class majority had been replaced by an African majority. And Africans continued to increase their relative share of the middle class between 2000 and 2008. By 2008 the number of middle-class Africans out-numbered middle-class whites by roughly two to one. This is a complete reversal of the demographic profile of the middle class from 1993.

The large change in the African-to-white ratio of the middle class is explained mostly by a large rise in the number of middle-class Africans, but is also explained by a significant fall in the number of middle-class whites (by approximately 1.2 million individuals). Shrinking numbers of middle-class whites are partly caused by restructuring in the class distribution for whites towards the top. Amongst the white population, the percentage of whites who were in the middle class fell from 81% to 67% between 1993 and 2008, with a shift towards more whites in the upper class (increasing from 7.8% to 20%). However, a further important factor is a change at the population level – the white population shrunk in size between 1993 and 2008 by approximately one million. Shrinking white population growth can be attributed to a combination of low fertility levels and to white emigration (Van Rooyen, 2000). Overall, a dwindling white population combined with a distributional shift amongst whites away from the middle class (and towards the upper class) resulted in a fast declining white share (and a rising African share) of the middle class over the period. Growing numbers of middle-class Africans combined with shrinking numbers of middle-class whites therefore partly explain how the total growth of the middle class between 1993 and 2008 was muted.

The percentage of the middle class who were coloured or Indian remained more stable over the period, but also showed positive transformation) (Table 2, Figure 3). Amongst coloureds, an additional 530 000 individuals were added to the middle class between 1993 and 2008. This was against the backdrop of coloured population growth of 900 000 individuals over the same period. Hence, the percentage of coloureds who were middle class increased from 23% to 31% and the total share of coloureds in the middle class as a whole increased from 10% to 13% over the period. Less can be said about changes in the size and share of Indians in the middle class (partly owing to the small sample size for Indians).

The racial composition of the upper class also showed significant racial transformation over the period (although the upper class remains a very small share of the total

Table 2: Race and class status, 1993–2008

	Lower class (<R515)			Lower class (R515 to R1 399)			Middle class (R1 400 to R10 000)			Upper class (>R10 000)			Total		
	1993	2000	2008	1993	2000	2008	1993	2000	2008	1993	2000	2008	1993	2000	2008
African															
Count (×1 000)	21399	23053	23438	6755	7769	9361	2235	4006	5377	19	112	257	30407	34940	38434
	(96)	(89)	(221)	(77)	(66)	(202)	(48)	(56)	(177)	(5)	(17)	(47)	(84)	(99)	(305)
Row %	70.4	66.0	61.0	22.2	22.2	24.4	7.3	11.5	14.0	0.1	0.3	0.7	100	100	100
Column %	94.1	93.7	93.5	74.5	79.0	79.9	29.0	46.2	52.3	4.3	13.5	19.1	76.1	79.5	79.4
White															
Count (×1 000)	183	87	125	375	298	473	4158	3055	2958	400	650	888	5116	4089	4444
	(15)	(10)	(25)	(22)	(29)	(65)	(72)	(66)	(155)	(23)	(33)	(105)	(79)	(79)	(199)
Row %	3.6	2.1	2.8	7.3	7.3	10.6	81.3	74.7	66.6	7.8	15.9	20.0	100	100	100
Column %	0.8	0.4	0.5	4.1	3.0	4.0	54.0	35.2	28.8	92.4	78.5	65.9	12.8	9.3	9.2
Coloured															
Count (×1 000)	1054	1362	1338	1551	1401	1608	790	1062	1320	8	35	43	3403	3859	4308
	(33)	(26)	(64)	(40)	(25)	(74)	(29)	(24)	(93)	(3)	(6)	(14)	(58)	(43)	(134)
Row %	31.0	35.3	31.0	45.6	36.3	37.3	23.2	27.5	30.6	0.2	0.9	1.0	100	100	100
Column %	4.6	5.5	5.3	17.1	14.2	13.7	10.2	12.2	12.8	1.9	4.2	3.2	8.5	8.8	8.9
Indian															
Count (×1 000)	101	96	179	389	368	268	525	554	633	6	32	158	1020	1050	1238
	(9)	(8)	(46)	(18)	(20)	(42)	(28)	(18)	(69)	(2)	(5)	(48)	(35)	(29)	(105)
Row %	9.9	9.1	14.5	38.1	35.0	21.6	51.4	52.8	51.1	0.6	3.0	12.8	100	100	100
Column %	0.4	0.4	0.7	4.3	3.7	2.3	6.8	6.4	6.2	1.3	3.9	11.8	2.6	2.4	2.6

(Table continued)

Table 2: Continued

	Lower class (<R515)			Lower class (R515 to R1 399)			Middle class (R1 400 to R10 000)			Upper class (>R10 000)			Total		
	1993	2000	2008	1993	2000	2008	1993	2000	2008	1993	2000	2008	1993	2000	2008
Total															
Count (×1 000)	22737	24597	25080	9070	9 836	11710	7707	8677	10288	432	828	1347	39946	43938	48425
	(96)	(90)	(230)	(86)	(77)	(224)	(90)	(89)	(258)	(23)	(38)	(126)	(45)	(111)	(362)
Row %	56.9	56.0	51.8	22.7	22.4	24.2	19.3	19.7	21.2	1.1	1.9	2.8	100	100	100
Column %	100	100	100	100	100	100	100	100	100	100	100	100	100	100	100

Source: PSLSD 1993, IES/LFS 2000, NIDS 2008; author's own estimates.

Notes: Standard errors in parentheses; the data are weighted.

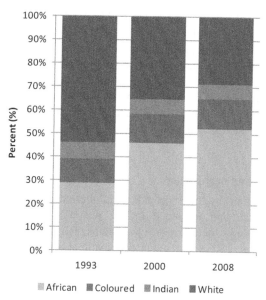

Figure 3: Racial composition of the middle class, 1993–2008
Source: PSLSD 1993, IES/LFS 2000, NIDS 2008.
Notes: Standard errors in parentheses; the data are weighted.

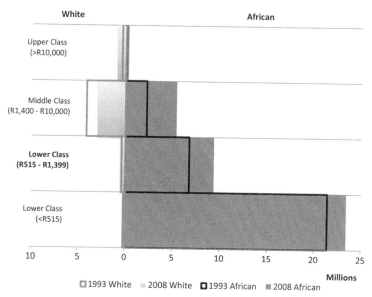

Figure 4: Class status of the African and White populations, 1993 and 2008
Source: PSLSD 1993, IES/LFS2000, NIDS 2008.
Notes: Standard errors in parentheses; the data are weighted.

population). In particular, there was a 10-fold increase in the number of upper-class Africans between 1993 and 2008, as opposed to a two-fold increase in upper-class whites. However, whites still held the racial majority of the upper class in 2008.

Large shifts in the racial composition of the middle (and upper) class are not repeated when examining the lower class. In fact, the significant feature of the racial composition of the lower classes is their homogeneous African composition. The racial share of individuals who fell below the R515 per-capita per-month poverty line fluctuated between 94.1 and 93.5% African, whilst the racial share of individuals who fell above the poverty line but were still in the lower class fluctuated between 74.5 and 79.9% African. Structural shifts in the racial composition of the class structure are therefore limited to affluence at the top of the distribution.

4.3 Black affluence and occupation

Investigating occupation by race provides an alternative means of verifying the transformation in race and affluence in the post-apartheid period (as generally followed in the sociology literature on class status; Rivero et al., 2003; Seekings & Nattrass, 2005; Muller, 2006). The major pragmatic shortcoming of the occupational approach to measuring class status is that it excludes a large proportion of the population who are not within the labour force, and it may also classify individuals of the same household (who arguably share resources and enjoy the same lifestyle) into different class categories. The use of household per-capita income by economists to define the middle class has the advantage of overcoming such obstacles. Irrespective, occupation and income are obviously correlated.

An investigation of race and class by occupation (see Table 3 and Figures 5, 6 and 7) shows that there has been a large change in the racial composition of typically 'middle class' occupations between 1993 and 2008.[4] The demographic composition of middle-class occupations shifted from majority white to majority African over the period (amongst clerks, Africans were already in the majority in 1993, but the African population share of clerks increased over the period).

These shifts were the result of a declining white share of total employment (due to the decline in the size of the total white population), and large absolute increases in the number of Africans in middle-class occupations (in excess of total African employment growth). Coloureds also experienced large growth in the number of middle-class employees, in excess of coloured employment growth (with the exception of clerical occupations). Interestingly, whites maintained the same number of jobs in professional occupations whilst losing a large number of management/ administrator positions. This may reflect the importance of the public sector in creating a large number of middle-class jobs for non-whites.

In sum, the racial composition of the middle class and upper class displayed massive transformation between 1993 and 2008 (as reflected in the income distribution and the occupational hierarchy).[5] That said, there is still significant progress to be made. Whites remain significantly over-represented in the middle (and upper) class relative

---

[4]The upper class is not ascribed a different set of occupations from that of the middle class in this analysis (see Visagie & Posel, 2013). The middle class is not always differentiated from the upper class in studies that rely on occupation to define class status (Rivero et al., 2003; Seekings & Nattrass, 2005).
[5]The specific size of the racial shift will depend on the choice of income boundary for the affluent middle class (analogous to the choice of poverty line). Irrespective of whether Africans actually outnumber whites in the middle class by 2008 (which will depend on the choice of threshold), there was a notable shift of Africans into higher income thresholds in much larger numbers than whites.

Table 3: Middle-class occupations by race, 1993–2008

	Mangers, senior officials and legislators			Professionals, associate professionals and technicians			Clerks			Total employed workforce		
	1993	2000	2008	1993	2000	2008	1993	2000	2008	1993	2000	2008
African												
Count (×1 000)	40 (6)	143 (11)	229 (35)	441 (21)	770 (28)	951 (67)	N/A	413 (17)	561 (49)	5 849 (70)	7 955 (70)	9 228 (194)
Row %	0.7	1.8	2.5	7.5	9.7	10.3	–	5.2	6.1	100	100	100
Column %	7.9	27.8	38.1	38.3	48.7	50.4	–	41.5	55.4	60.7	69.4	70.8
White												
Count (×1 000)	414 (23)	297 (20)	215 (40)	572 (27)	602 (37)	632 (80)	N/A	365 (21)	232 (46)	2 331 (54)	1 836 (55)	1 876 (129)
Row %	17.8	16.2	11.5	24.6	32.8	33.7	–	19.9	12.4	100	100	100
Column %	80.7	57.8	35.8	49.6	38.1	33.5	–	36.6	22.9	24.2	16.0	14.4
Coloured												
Count (×1 000)	29 (6)	30 (4)	67 (17)	77 (9)	129 (9)	207 (28)	N/A	147 (9)	138 (28)	1 082 (34)	1 282 (25)	1 415 (79)
Row %	2.7	2.4	4.8	7.1	10.1	14.6	–	11.4	9.8	100	100	100
Column %	5.6	5.9	11.2	6.7	8.2	11.0	–	14.8	13.7	11.2	11.2	10.9
India												
Count (×1 000)	30 (5)	44 (5)	89 (32)	63 (11)	80 (7)	98 (30)	N/A	71 (7)	80 (31)	368 (22)	385 (21)	515 (74)
Row %	8.1	11.4	17.4	17.1	20.7	19.0	–	18.3	15.6	100	100	100
Column %	5.8	8.5	14.9	5.5	5.0	5.2	–	7.1	7.9	3.8	3.4	3.9
Total												
Count (×1 000)	513 (25)	514 (24)	600 (64)	1 153 (37)	1 581 (47)	1 888 (112)	N/A	995 (29)	1 011 (79)	9 631 (93)	11 458 (93)	13 035 (253)
Row %	5.3	4.5	4.6	12.0	13.8	14.5	–	8.7	7.8	100	100	100
Column %	100	100	100	100	100	100	–	100	100	100	100	100

Source: PSLSD 1993, IES/LFS 2000, NIDS 2008; author's own estimates.

Notes: Standard errors in parentheses; the data are weighted.

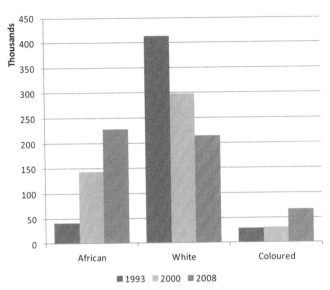

Figure 5: Managers, legislators and senior officials by race, 1993–2008
Source: PSLSD 1993, IES/LFS 2000, NIDS 2008.
Notes: The data are weighted.

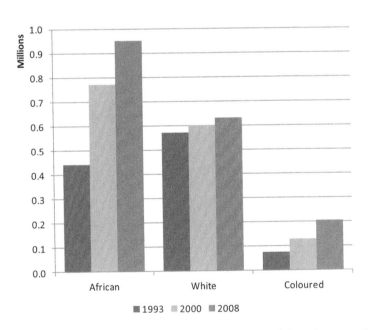

Figure 6: Professionals, associate professional and technicians by race, 1993–2008
Source: PSLSD 1993, IES/LFS 2000, NIDS 2008.
Notes: The data are weighted.

to their population share (and under-represented in the lower classes), whereas Africans are significantly under-represented in the middle class in relation to their population share.

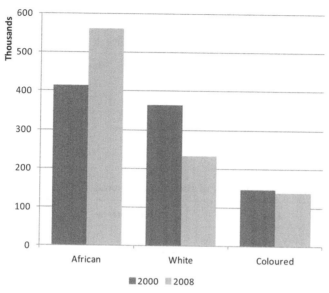

Figure 7: Clerks by race, 1993–2008
Source: PSLSD 1993, IES/LFS 2000, NIDS 2008.
Notes: The data are weighted.

Moreover, the preceding analysis of the middle class says little about changes in livelihoods for the actual middle. Although racial equity at the top of the income distribution is welcomed, of arguably equal (if not greater) importance is the economic empowerment of the majority. This leads to our second perspective of 'middle class' development – an examination of changes in incomes for the middle-income strata between 1993 and 2008.

5. Perspective 2: The actual middle and depressed income growth

5.1 The middle-income squeeze

Table 4 presents the median per-capita income and corresponding income thresholds (50 to 150% of the median) for individuals residing in households that fell in the actual middle of the distribution of income between 1993 and 2008. In 1993 the middle strata received between R345 and R1034 per capita per month (in constant 2008 prices). This is very revealing of the economic hardships faced by the majority of South Africans at the start of the democratic era. In fact, approximately 40% of individuals located in the middle strata fell below the poverty line of R515 per capita per month. The actual middle are anything but 'middle-class' when compared with the standard of living associated with middle-class affluence.

Incomes were very slow to rise between 1993 and 2008. The median per-capita household income increased from R689 per capita per month in 1993 to R749 in 2000 and ended on R759 in 2008. This represents a modest 10% increase in the median per-capita household income over the 15-year period of review. This translates into an average per-capita growth rate of just 0.67% per annum.

Table 5 presents information on the size and income shares of the lower, middle and upper strata for the period. In absolute terms, total income collected by households in the middle increased from R7.1 billion in 1993 to R9.9 billion in 2008. However, the

Table 4: Thresholds for the middle-income strata, 1993–2008

	1993 (R)	2000 (R)	2008 (R)
Median	689	749	759
Lower bound	345	374	380
Upper bound	1 034	1 123	1 139

Source: PSLSD 1993, IES/LFS 2000, NIDS 2008.

Notes: The data are weighted.

Table 5: Size and income shares of the income strata, 1993–2008

	Lower strata			Middle strata			Upper strata		
	1993	2000	2008	1993	2000	2008	1993	2000	2008
Class size									
Count (millions)	2.9	3.6	4.2	2.5	3.4	4.2	3.6	4.2	5.2
	(0.1)	(0.0)	(0.1)	(0.0)	(0.0)	(0.1)	(0.1)	(0.1)	(0.2)
Percentage share	32.3	32.1	30.9	27.7	30.2	30.9	40.1	37.6	38.2
	(0.5)	(0.4)	(0.8)	(0.5)	(0.4)	(0.8)	(0.5)	(0.5)	(1.0)
Total income from all sources									
Count (billions)	R2.9	R3.9	R3.9	R7.1	R8.2	R9.9	R36.5	R48.2	R63.6
	(0.0)	(0.0)	(0.0)	(0.1)	(0.1)	(0.2)	(0.7)	(0.8)	(2.7)
Percentage share	6.3	6.4	5.0	15.2	13.6	12.7	78.5	80.1	82.3

Source: PSLSD 1993, IES/LFS 2000, NIDS 2008; author's own estimates.

Notes: Standard errors in parentheses; the data are weighted.

relative share of income accruing to these households actually declined from 15.2% to 12.7% of total income between 1993 and 2008. This is in spite of growth in the relative size of the middle strata (in terms of the share of households) from 27.7% to 30.9% of households over the period. The lower strata similarly experienced a fall in their share of total income.

Of course it follows that relatively less income received by the lower and middle strata means more income received by households in the upper strata. In fact, it was households at the very top of the upper strata that increased their relative income share. Households in the 10th (and most affluent) income decile increased their share of income from 40.7% to 48.9% of total income[6] (this is also shown by Leibbrandt et al., 2010).

Therefore it follows that the middle strata did not experience 'inclusive' growth over the period. The growth incidence curve shown in Figure 8 illustrates this point visually (see Ravallion & Chen, 2003). For economic growth to be 'inclusive', the curve should angle downwards from left to right; that is, growth should be highest over the lower deciles and

---

[6]Given the extent of income inequality in South Africa, the upper bound of 150% of the median income results in a relative large upper strata. Considering the top 10% of households or individuals within the distribution provides further clarity as to the income status of those at the very top of the income distribution.

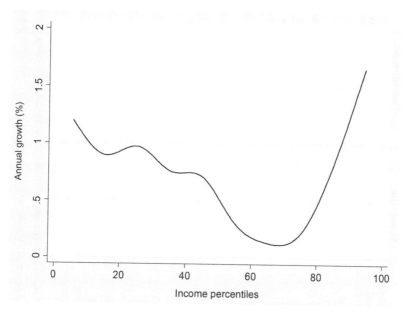

Figure 8: Growth incidence curve: annualised per capita income growth, 1993–2008

Source: PSLSD 1993, NIDS 2008; author's own estimates.

Notes: The data are weighted.

lowest over the upper deciles. Although annualised per-capita growth was consistently positive between 1993 and 2008, it was at its lowest over the middle deciles. As seen in the figure, the growth incidence curve is roughly 'U-shaped', suggesting a polarised pattern of income growth, but strongly biased in the upper tail of the curve. Relatively better income growth in the lower tail of the curve (in comparison with the middle) reflects the impact of the large expansion in government social grants over the period (Bhorat & Van der Westhuizen, 2008; Van der Berg et al., 2008). This U-shaped pattern of growth may partially explain how falling poverty levels have gone hand-in-hand with rising income inequality in post-apartheid South Africa.

5.2 Labour market earnings and social grants

To better understand this middle-income squeeze, Table 6 (see also Figure 9) provides a breakdown of the composition of total household income and how this changed over the period.

The middle strata became less reliant upon earnings from the labour market, both in terms of the average rand value of earnings received (constant 2008 prices) and even more so in percentage terms. In 1993, 70% of total income was derived from the labour market, whereas in 2008 only 59% of total income came through the labour market. The lower strata similarly experienced a fall in labour market earnings, although not of the same magnitude as households in the middle strata.

Eroding labour market incomes post 1994 are the result of significant structural shifts in the labour market well documented in the literature. These changes include: falling levels of formal sector employment in the 1990s (Seekings & Nattrass, 2005); a general shift

Table 6: Sources of household per capita income, 1993–2008

		Lower strata			Middle strata			Upper strata			Population total		
		1993	2000	2008	1993	2000	2008	1993	2000	2008	1993	2000	2008
Labour market earnings	Mean Rands per capita	74	88	62	449	453	408	3361	4240	4131	1 496	1761	1724
		(2)	(1)	(3)	(6)	(4)	(11)	(98)	(111)	(209)	(43)	(47)	(91)
	Column %	40.1	41.7	31.9	70.0	66.5	59.2	88.8	86.5	85.0	68.6	66.2	61.8
Welfare grants	Mean Rands per capita	55	47	91	111	185	187	34	66	58	62	66	108
		(2)	(1)	(3)	(5)	(2)	(8)	(4)	(7)	(7)	(2)	(3)	(4)
	Column %	25.8	23.1	55.3	16.8	13.6	29.7	1.7	2.2	2.9	13.4	12.3	26.2
UIF and workers compensation	Mean Rands per capita	3	17	1	9	38	7	16	123	10	10	63	7
		(0)	(0)	(1)	(1)	(1)	(4)	(4)	(11)	(5)	(2)	(4)	(2)
	Column %	1.9	10.5	0.7	1.5	5.8	1.0	0.5	3.3	0.3	1.2	6.3	0.6
Investments	Mean Rands per capita	2	4	2	14	15	10	423	302	499	174	119	194
		(0)	(0)	(1)	(2)	(1)	(3)	(39)	(28)	(78)	(16)	(10)	(30)
	Column %	1.5	1.9	1.1	2.2	2.2	1.6	7.2	5.3	7.8	4.0	3.2	3.9
Remittances	Mean Rands per capita	40	43	19	53	73	53	39	87	208	43	68	102
		(1)	(1)	(1)	(3)	(3)	(6)	(4)	(10)	(78)	(2)	(4)	(30)
	Column %	30.8	22.8	11.1	9.5	11.9	8.6	1.7	2.8	4.0	12.8	11.9	7.5
Total	Mean Rands per capita	173	199	175	635	664	666	3873	4818	4907	1785	2078	2 135
		(2)	(1)	(3)	(4)	(3)	(8)	(103)	(114)	(224)	(46)	(49)	(107)
	Column %	100.0	100.0	100.0	100.0	100.0	100.0	100.0	100.0	100.0	100.0	100.0	100.0

Source: PSLSD 1993, IES/LFS200, NIDS 2008; author's own estimates.

Notes: The data are weighted. Some households did not provide a detailed breakdown of all their sources of income and hence could not be included in the estimates. This means that population totals in the table do not correspond with the population totals presented elsewhere. Of concern is the lack of income growth amongst the lower strata, which should show larger income growth in accordance with Figure 8. Despite these concerns, changes in the pattern of income growth are very large, and selection bias is unlikely to explain all of the shifts in the sources of income. Unemployment Insurance Fund (UIF).

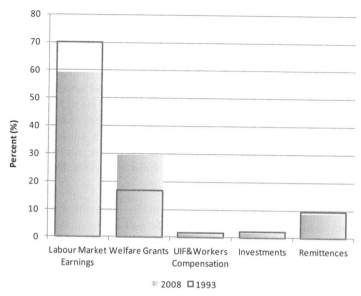

Figure 9: Changes in the sources of income of the middle-income strata, 1993 and 2008

Source: PSLSD 1993, NIDS 2008; author's own estimates.

Notes: The data are weighted. UIF, Unemployment Insurance Fund.

from unskilled to skilled labour employment (Oosthuizen, 2003; Bhorat, 2005; Rodrik, 2006); a large increase in labour supply ahead of employment growth (Casale et al., 2004); and resulting steady increases in the unemployment rate (whether strict or expanded) (Kingdon & Knight, 2004, 2007; Seekings & Nattrass, 2005). All of the abovementioned trends undermine the aggregate labour market earnings of households in the lower and middle strata in South Africa.

However, falling labour market earnings both amongst the lower and middle strata did not lead to a net reduction in per-capita income. This was due to a large increase in government grant income. Amongst the lower income strata, government grant income increased from just 26% of total income in 1993 to the majority share at 55% of total income in 2008. Even in the middle, the average percentage of income derived from government grants for middle-income households increased from 17% to 30% between 1993 and 2008. Both the lower strata and the middle strata were therefore cushioned from a regression in household incomes through fiscal redistribution. However, the South African fiscus will not allow for the same level of expansion in social welfare into the future – at least not without an ideological change regarding the acceptable size of the budget deficit.

6. Policy implications

The above two perspectives of the 'middle class' have important implications for economic policy. In terms of middle-class affluence, there is strong evidence of progress in the racial composition of the middle (and upper) class. Although race is still highly correlated with class, the relationship has moved in the right direction. In fact, use of my particular definition of the middle class finds a switch from majority white to majority African between 1993 and 2008 – and this is also borne out in the racial mix of middle-class occupations. Hence, South Africa has moved closer towards racial equity in the class structure as targeted in Affirmative Action and BEE policy.

However, inequality in the South African distribution of income has not fundamentally changed, and has even worsened. Income became even more concentrated at the very top. South Africa therefore moved away from race-based cleavages towards class-based cleavages, and such class-based cleavages strengthened over the period. The upper class experienced very large growth (albeit from a low base), whereas the affluent middle grew only slowly. Policy needs to balance out the need for racial transformation at the top of the income hierarchy alongside the need for deeper structural reform of the class system itself.

A focus on households in the 'actual middle' of the distribution shows very modest income growth between 1993 and 2008 of only half a percentage point per annum. This is the result of lost traction in the labour market. If it was not for massive expansion and take-up of social grants in South Africa, incomes for the lower and middle strata may have regressed. Such lacklustre income progress at the middle does not bode well for future social stability. South Africa needs to actualise a far more inclusive growth trajectory – current performance reflects a failure of macroeconomic policy in creating jobs and significantly raising livelihoods for the majority.

Acknowledgements

The author would like to thank Economic Research Southern Africa (ERSA) for their support of this project as well as to thank the editor from the Econ3x3 Forum for useful feedback on this topic.

References

African Development Bank, 2011. The middle of the pyramid: dynamics of the middle class in Africa. Market Brief. Chief Economist Complex, Tunis-Belvedûre, Tunisia.

Atkinson, A & Brandolini, A, 2013. On the identification of the middle class. In Gornick, J & Jantti, M (Eds), Income Inequality: Economic Disparities and the Middle Class in Affluent Countries. Stanford University Press, Stanford, CA.

Barro, R, 1999. Determinants of democracy. Journal of Political Economy 107(6), 158–83.

Bhalla, S, 2007. Second Among Equals: The Middle Class Kingdoms of India and China. Peterson Institute for International Economics, Washington, DC.

Bhorat, H, 2005. Labour market challenges in the post-apartheid South Africa. South African Journal of Economics 72(5), 940–77.

Bhorat, H & van der Westhuizen, C, 2008. The Regulatory Environment and its Impact on the Nature and Level of Economic Growth and Development in South Africa, DPRU Conference, 27–29 October, Glenburn Lodge, Muldersdrift, South Africa.

Birdsall, N, 2010. The (indispensable) middle class in developing countries. In Kanbur, R & Spence, M (Eds), Equity and Growth in a Globalizing World. The World Bank, Washington, DC.

Brown, C, 2004. Does income distribution matter for effective demand? Evidence from the United States. Review of Political Economy 16(3), 291–307.

Casale, D, Muller, C & Posel, D, 2004. Two million net new jobs: A reconsideration of the rise in employment in South Africa, 1995–2003. South African Journal of Economics 72(5), 978–1002.

Davis, J & Huston, J, 1992. The shrinking middle-income class: A multivariate analysis. Eastern Economic Journal 18(3), 277–85.

Doepke, M & Zilibotti, F, 2005. Social class and the spirit of capitalism. Journal of the European Economic Association 3(2–3), 516–524.

Easterly, W, 2001. The middle class consensus and economic development. Journal of Economic Growth 6, 317–35.

Easterly, W, 2007. Inequality does cause underdevelopment: Insights from a new instrument. Journal of Development Economics 84, 755–76.

Finn, A, Leibbrandt, M & Levinsohn, J, 2012. Income mobility in South Africa: Evidence from the first two waves of the National Income Dynamics Study. SALDRU Working Paper Number 82/NIDS Discussion Paper 2012/5, Southern Africa Labour and Development Policy Research Unit (SALDRU), Cape Town.

Galor, O & Zeira, J, 1993. Income distribution and macroeconomics. Review of Economic Studies 60, 35–52.

Hoffman, E, 2008. A wolf in sheep's clothing: Discrimination against the majority undermines equality while continuing to benefit few under the guise of Black Economic Empowerment. Syracuse Journal of International Law and Commerce 36, 87–115.

Hoogeveen, J & Özler, B, 2006. Not separate, not equal: Poverty and inequality in post-apartheid South Africa. In Bhorat, H & Kanbur, R (Eds), Poverty and Policy in Post-apartheid South Africa. HSRC Press, Cape Town.

Kharas, H & Gertz, G, 2010. The new global middle class: A cross-over from West to East. In Li, C (Ed.), China's Emerging Middle Class: Beyond Economic Transformation. Brookings Institute Press, Washington, DC.

Kingdon, G & Knight, J, 2004. Unemployment in South Africa: The nature of the beast. World Development 32(3), 391–408.

Kingdon, G & Knight, J, 2007. Unemployment in South Africa, 1995–2003: Causes, problems and policies. Journal of African Economies 16(5), 813–48.

Landes, D, 1998. The Wealth and Poverty of Nations. Norton, New York.

Leibbrandt, M, Woolard, I, Finn, A & Argent, J, 2010. Trends in South African income distribution and poverty since the fall of apartheid. Organisation for Economic Co-operation and Development Social, Employment and Migration Working Papers 101, OECD, Paris.

Marx, K, 1974. Capital. Lawrence and Wishart, London.

Muller, S, 2006. The challenge of joining two economies: Assessing changes in the black middle class post-1994. In Gqubule, D (Ed.), Making Mistakes Righting Wrongs: Insights into Black Economic Empowerment. Jonathan Ball, Johannesburg.

Oosthuizen, M, 2003. Expected labour demand in South Africa: 1998–2003. Development Policy Research Unit Working Paper 03/81, DPRU, Cape Town.

Palma, J, 2011. Homogeneous middles vs. heterogeneous tails, and the end of the 'inverted-U': It's all about the share of the rich. Development and Change 42(1), 87–153.

Perotti, R, 1996. Growth, income distribution, and democracy: What the data say. Journal of Economic Growth 1, 149–88.

Ponte, S, Roberts, S & van Sittert, L, 2007. Black economic empowerment, business and the state in South Africa. Development and Change 38(5), 933–55.

Posel, D & Rogan, M, 2009. Women, income and poverty: Gendered access to resources in post-apartheid South Africa. Agenda 81, 25–34.

Pressman, S, 2007. The decline of the middle class: An international perspective. Journal of Economic Issues 41(1), 181–200.

Ravallion, M, 2010. The developing world's bulging (but vulnerable) middle class. World Development 38(4), 445–54.

Ravallion, M & Chen, S, 2003. Measuring pro-poor growth. Economic Letters 78, 93–9.

RDP (Reconstruction and Development Programme), 1994. White Paper on Reconstruction and Development Programme. Government Gazette vol 353, no. 16085. Notice no. 1954 of 1994, Parliament of The Republic of South Africa, Cape Town.

Rivero, C, Du Toit, P & Kotze, H, 2003. Tracking the development of the middle class in democratic South Africa. Politeia 22(3), 6–29.

Rodrik, D, 2006. Understanding South Africa's economic puzzles. CID Working Paper No. 130, Center for International Development at Harvard University, Cambridge, MA.

Schlemmer, L, 2005. Lost in transformation? South Africa's emerging middle class. Centre for Development and Enterprise Focus Papers 8, Centre for Development and Enterprise, Johannesburg.

Seekings, J & Nattrass, N, 2005. Class, Race and Inequality in South Africa. UKZN Press, Durban.

Sokoloff, K & Engerman, S, 2000. Institutions, factor endowments, and path of development in the New World. Journal of Economic Perspectives 14(3), 217–32.

Southall, R, 2007. Ten propositions about black economic empowerment in South Africa. Review of African Political Economy 34 (111), 67–84.

Thurow, L, 1987. A surge in inequality. Scientific American 256, 30–7.

Van der Berg, S, 2010. Straddling Two Worlds: Income Inequality in South Africa. Centre for Development and Enterprise, Johannesburg.

Van der Berg, S & Louw, M, 2004. Changing patterns of South African income distribution: Towards time series estimates of distribution and poverty. South African Journal of Economics 72(3), 546–72.

Van der Berg, S, Louw, M & Yu, D, 2008. Post-transition poverty trends based on an alternative data source. South African Journal of Economics 76 (1), 59–76.

Van Rooyen, J. 2000. The New Great Trek: The Story of South Africa's White Exodus. Unisa Press, Pretoria, South Africa.

Visagie, J & Posel, D, 2013. A reconsideration of what and who is middle class in South Africa. Development Southern Africa. DOI:10.1080/0376835X.2013.797224

Weber, M, 1961. Essays in Sociology. Routledge and Kegan Paul, London.

Whiteford, A & Van Seventer, D, 2000. South Africa's changing income distribution in the 1990s. Studies in Economics and Econometrics 24(3), 7–30.

The emergent middle class in contemporary South Africa: Examining and comparing rival approaches

Ronelle Burger[1], Cindy Lee Steenekamp[2], Servaas van der Berg[3] & Asmus Zoch[4]

In light of the economic, political and social significance of the middle class for South Africa's emerging democracy, we critically examine contrasting conceptualisations of social class. We compare four rival approaches to empirical estimation of class: an occupational skill measure, a vulnerability indictor, an income polarisation approach and subjective social status. There is considerable variation in who is classified as middle class based on the definition that is employed and, in particular, a marked difference between subjective and objective notions of social class. We caution against overoptimistic predictions based on the growth of the black middle class. While the surge in the black middle class is expected to help dismantle the association between race and class in South Africa, the analysis suggests that notions of identity may adjust more slowly to these new realities and consequently racial integration and social cohesion may emerge with a substantial lag.

1. Introduction

Apartheid has left its scar on South Africa's social landscape. Almost two decades after the first fully inclusive elections in 1994, society continues to be characterised by a lack of social cohesion and economic injustice due to the persistence of race as a marker and a sorting mechanism in many dimensions of daily life, including geographical space, educational opportunities, the labour market, social networks and political party affiliation.

However, the post-apartheid period has also seen rapid growth of the middle class and a further expansion of the black middle class, building on the rapid increase between 1969 and 1983 documented by Crankshaw (1997). These developments have been interpreted as signs that race may be gradually be becoming disassociated from class. A growing and more racially representative middle class can help shape a more just, dynamic and integrated society.

Theory and previous empirical findings suggest that an expanding middle class can boost economic efficiency and growth, can enhance the effectiveness and the stability of democratic institutions and political processes, and can help to mend social

[1]Researcher, Department of Economics, Stellenbosch University, Private Bag X1, Matieland 7602, South Africa.
[2]Research Fellow, Centre for International and Comparative Politics, Department of Political Science, Stellenbosch University, Private Bag X1, Matieland 7602, South Africa.
[3]Professor, Department of Economics, Stellenbosch University, Private Bag X1, Matieland 7602, South Africa.
[4]Researcher, Department of Economics, Stellenbosch University, Private Bag X1, Matieland 7602, South Africa.

fragmentation and polarisation. There are also arguments that a growing middle class can have a direct impact on social cohesion, seemingly based on the intuition that a larger middle group would often imply lower polarisation and would serve as a buffer between the rich and the poor.

Within the South African context it is of course significant that a large share of the new middle-class members is non-white: both from a retrospective viewpoint, as evidence that some of the post-apartheid reforms have been successful in creating a more just and dynamic society, but also from a prospective viewpoint, as a catalyst for further social change.

Owing to the economic, political and social significance of class for South Africa's emerging democracy, it is important and interesting to estimate the magnitude of the growth of the middle class, and specifically the black middle class. However, such attempts are frustrated by disputes around the origins and meaning of class and how it should be measured. Lopez-Calva & Ortiz-Juarez (2011a:1) highlight that empirical work on the middle class tends to assume that 'thresholds that define relatively homogeneous groups in terms of pre-determined sociological characteristics can be found empirically'. Research on the middle class is often motivated by an interest in the political, economic and social benefits associated with the term, but without verifying whether the selected empirical approach is aligned with such conjectured benefits.

Using hyperbole and humour to highlight the proliferation of competing conceptualisations of the middle class, Beckett (2010:1) refers to it as 'a slippery business' that has in the past been associated with a long and divergent list of characteristics, including 'having servants, renting a good property, owning a good property, owning a business, being employed in one of the professions, how you speak, how you use cutlery' (2010:1). This article acknowledges the contradictions and tensions in the menu of definitions and methods available for approximating the size of the middle class. It pursues a more critical and conceptually grounded approach. The next section presents a brief, but critical overview of contrasting conceptualisations of class and how this has affected studies on class in South Africa.

2. Estimating the size of the middle class in South Africa

The analysis of class is rooted in the pioneering work of Karl Marx and Max Weber (Marx, 1867; Marx & Engels, 1968; Weber, 1968). Marx defines class as shared structural positions within the social organisation of production. Class originates from shared interests and economic positions. The focus is on a conflict between the two main social classes: bourgeoisie (upper class) and proletariat (lower class). The bourgeoisie owns or controls the means of production (physical capital) while the proletariat does not and therefore needs to sell their labour to the bourgeoisie. Marx also distinguished a third class of petty bourgeoisie, which can be seen as a middle class. They are typically small business owners, shop-keepers, artisans and managers who are similar to the bourgeoisie because they are able to control (but not necessarily own) the means of production, but differ from the haute bourgeoisie because they work alongside their staff. Marx saw this group as a transitional class, which would eventually be absorbed into the proletariat.

In contrast, Max Weber saw shared life chances as the basis of class. Life chances were associated with opportunities for generating income in the market. Similar to Marx he also differentiates between those with access to property and land and those without, who have to earn their living through work. For those without property, their education, skills and knowledge determine their market value, their occupation and their wages, and in turn wages determine the lifestyle that an individual can afford. In this way, economic position maps to social status and shapes shared interests and social communities. However, while economic power, social status and political power are often correlated, Weber diverges from Marx by recognising that social status and political power are separate dimensions that are not always aligned with economic position and power. According to Weber, differences in social status can emerge due to ownership of the means of production, but it can also emerge due to other factors such as skills or credentials (Seekings, 2009).

Weber also distinguished the established class of property owners from the emergent middle class who were white-collar employees without property. However, in contrast to Marx, who saw this class as a temporary or transitional class that will eventually be absorbed into the proletariat, Weber expected the increasing bureaucratisation of administration to enhance the importance of specialist examination, creating a 'universal clamour for the creation of educational certificates in all fields' and leading to 'the formation of a privileged stratum in bureaus and offices' (1961:241).

The concept of class has evolved much since the days of Karl Marx and Max Weber, but education, social status, income, wealth and shared life perspectives have remained central to definitions of class. One of the main enduring tensions is whether education or income is at the heart of the definition of class and the main transmission mechanism for the benefits of a growing middle class (Mattes, forthcoming).

Despite debates around the relevance of the concept of class[5] in current times, the term 'middle class' has continued to be popular amongst both researchers and the media. Recently, it has often been used to gauge the pace of social change and economic advancement in emerging and developing economies. In this literature, the term 'middle class' is frequently used as shorthand for increased agency and empowerment that allow individuals to competently navigate their own destinies and realise their own potential.

The term's enduring popularity and continued prominence appear to be partly due to a significant literature linking the middle class to a range of desirable country-level outcomes such as social cohesion, political stability and economic growth. Although authors are seldom explicit about the transmission mechanisms for such benefits, several analytical linkages have been frequently mentioned, including appeasement of the poor, increasing discretion in the use of money and time as basic needs are met, and a longer planning horizon due to greater stability in living standards. The combination of more discretionary income and a longer time horizon is expected to encourage investment in physical and human capital – both of which are traditionally viewed as important for stimulating economic growth because they improve productivity – and to enable the middle class to be more active and vocal in promoting accountability, good governance and the prioritisation of public goods such as education (Birdsall, 2010). Mattes (forthcoming) distinguishes three different theories about how a larger middle class can strengthen democracy: through a more

---

[5]For instance, Pakulski (2005) argues that the complex configurations of classless inequality and antagonism call for more comprehensive theoretical and analytic constructs.

educated and skilled public-sector workforce; via the relationship between education and civic values; or by allowing a focus on free speech, civil liberties and democracy when basic needs have been met.

The interest in South Africa's social structure in general, and the middle class in particular, has spawned a large literature, much of which has had a strong Marxist focus. Seekings (2009) points out that there were a number of Weberian scholars working on class in South Africa between the late 1940s and the early 1970s, but that this stream of work was abandoned and subsequently forgotten due to apartheid-era political considerations and sentiments that favoured a Marxist approach. The dominance of the Marxist approach led to the neglect of the relationship between class and Weberian concepts such as skill and social status. Although not explicitly Weberian, more recent research tends to highlight the role of Weberian concepts such as education and consumption (e.g. Rivero et al., 2003; Schlemmer, 2005; Seekings & Nattrass, 2005; Nieftagodien & Van der Berg, 2007; Seekings, 2007; Visagie & Posel, 2013).

There is consensus amongst the more recent studies that the middle class is expanding and that there has been a significant increase in the black share of the middle class (e.g. StatsSA, 2009; Van der Berg, 2010). While this growing disassociation between race and class has the potential to promote political stability and social cohesion, Schlemmer (2005) concludes that the racial divide is still conspicuous – especially in areas where racial interests diverge, such as party politics, affirmative action and privatisation. He finds few close personal links across the racial divide. He also reports that there is no cohesive identity and coherence amongst the emergent black middle class. Many appear uncomfortable with the label and are reluctant to identify themselves as middle class.

Our analysis attempts to add to this existing literature by comparing rival conceptualisations of the middle class and assessing the overlaps and tensions between these definitions when applied to recent and representative South African datasets. In particular, in light of Schlemmer's (2005) findings, we examine how well traditional externally defined measures of class align with individuals' subjective notions of their social position and class identity.

3. Data

Our analysis relies mainly on the 2008 National Income Dynamics Study (NIDS) data, including approximately 7000 households. The survey contains detailed information on occupation, income and expenditure, assets and self-assessed social standing. We include the Project for Statistics on Living Standards and Development (PSLSD) of 1993 in the trend analysis where it was sufficiently comparable with the data from the three NIDS surveys.[6]

The second main data source is the last four waves (1995, 2001, 2006, 2013) of the World Values Survey (WVS). Each of the waves of the survey has a sample of approximately 3000 individual observations. This survey contains information on income brackets, occupations and assets and is unique due to questions on self-reported class, values, and attitudes.

---

[6]The survey includes approximately 9000 households.

Table 1: Characteristics of classes when categorised based on occupational skill levels, 2008

Class	Share of total Population (%)	Share of black population (%)	Share of white population (%)	Black share of each class (%)	Mean Income (Rands)	Mean Age (Years)	Mean educational attainment (Years)
HH not economically active	14.1	15	14.4	84.4	1060	43.5	7.2
Unemployed	16.1	18.5	3.3	91.2	551	33.4	8.9
Working & occ info missing	25.1	26.1	23.8	80.2	1803	39.3	8.7
Low skilled	10.1	10.9	1.8	86.3	902	38.4	7.1
Medium skilled	21.7	20.8	20.2	76.1	2001	36.0	10.0
Highly skilled	13.0	8.8	36.6	54.0	4919	39.0	12.8

Source: NIDS 2008. Household classifications are based on the highest skill level within the household. The mean age and education attainment relates to the household member with the highest occupational skill level.

4. Comparing rival approaches to estimating the middle class

We describe and implement the four main empirical approaches to defining the middle class – by skill or occupation, by vulnerability, by income and by self-identification – using the 2008 NIDS. This analysis is supplemented by work on the WVS, considering patterns in self-identification as middle-class members.

4.1 Occupation and skill level

There is a longstanding tradition of class analysis based on occupation and skill level (see for example Goldthorpe, 1987; Wright, 1980; Goldthorpe & Erikson, 1993; Evans & Millan, 1999) and consequently a fair degree of consensus has emerged regarding what occupations constitute the middle class (e.g. white-collar workers such as professionals, managers and clerks).

Internationally there are disputes about whether the middle class should include the self-employed (Wright, 1979, 1989; Glassman et al., 1993). In South Africa this question is also relevant because self-employment is often regarded as a last resort for those who cannot find salaried employment. The self-employed may therefore not be an elite group (as a Marxian analysis will often assume), but will include both highly skilled and unskilled individuals.

Furthermore, there are also concerns about accurately capturing and reflecting occupation and acquired skill. In settings with high unemployment rates and significant levels of underemployment, there may be a divergence between the productive characteristics and abilities of individuals and their occupation (Seekings & Nattrass, 2005). In some surveys there may be problems around reliable self-classification and missing values for occupational data.

A traditional shortcoming of occupational analysis is that the approach is unable to assign a class to any households with no employed household members, which is problematic in developing countries where this segment of the population may be a significant share. Seekings (2003) proposes that in South Africa the unemployed may constitute a separate class due to the size of this group and the large divide in the living conditions and life chances of the unemployed and the employed.[7] The frequent traffic out of unemployment and into low-skilled work and vice versa represents an obstacle to interpreting the unemployed as a clear and stable class category, distinct from the low-skilled occupational category. Additionally, classifying the unemployed as an underclass still leaves households with no economically active household members without any classification.

We navigate our way through the issues outlined above by implementing a definition of occupational class that has four broad classifications: unemployed, low skill, medium skill and high skill. The low-skill category comprises the so-called elementary occupations and includes domestic, agricultural and fishery workers. The medium-skill category is defined as clerks, service workers and shop and market, craft and related trades workers, plant and machinery operators and assemblers, while the high-skill category includes legislators, senior officials and managers, professionals, technicians and associated professions. The self-employed are classified based on their occupations, in the same way as regular employees. We convert individual-level occupational skill variables to a household variable by using the highest occupational skill level of any household member to classify households. This approach could be defended via Ceruti's (2013) perspective that individuals' class position could be mediated via the life chances and living standards of those with whom they share a home.

Omissions due to working household members not providing any occupational data or a lack of economically active household members constituted 25% and 14% of the population respectively in the NIDS 2008. Despite these shortcomings, the occupation variable is sufficiently reliable to provide a broad overview of the prevalence of middle-class occupations and its interaction with other variables of interest such as race, age and education.

Table 1 illustrates that the low-skilled and medium-skilled occupational classes combined represent 32% of all South African households. The low-skilled occupational class represents roughly 10%, the medium-skilled occupational class 21.7% and the two tails of the distribution (the highly skilled and the unemployed) 13% and 16% of the population, respectively. In total about 70% of households have at least one member working. Table 1 also shows that higher skilled occupations are associated with higher income levels and educational attainment. Blacks constitute 91% of the unemployed, 86% of households associated with lower skill occupations, 76% of households with medium-skill occupations and 54% of households with high-skilled occupations.

4.2 Vulnerability

Goldthorpe & McKnight (2004) find that class is correlated with the risk and uncertainty faced by individuals, mediated via the relationship between secure employment

---

[7]Seekings & Nattrass (2005) point out that there can be variation in the labour market prospects among the unemployed, including, for instance, skilled individuals from affluent households using unemployment as a waiting bay for the right opportunity. However, less than 1% of those who were unemployed in 2008 earned R10 000 or more in wages (measured in 2008 SA Rands) in the two consecutive waves of NIDS (2010, 2012).

Table 2: Characteristics of classes when categorised based on vulnerability approach, 2008

Class	Share of total population (%)	Share of black population (%)	Share of white population (%)	Black share of each class (%)	Mean income (Rands)	Mean age	Mean educational attainment (years)
Lower class	22.6	28.3	–	98.1	323	38.1	6.4
Vulnerable class	37.5	45.5	0.2	95.2	639	38.2	7.5
Middle class	26.8	23.8	22.6	69.7	1 925	36.5	9.9
Upper class	13.1	2.4	77.3	14.6	7 308	43.0	12.7

Source: NIDS 2008 and 2010. Based on the predicted probability of a household to be non-poor in 2010, households are classified as lower class (probability <10%); vulnerable (10 to 50%); middle class (50 to 90%) or upper class (>90%). The mean age and education attainment relate to the household member with the highest occupational skill level.

contracts, labour market negotiating power and skill scarcity. Following from this, Lopez-Calva & Ortiz-Juarez (2011a) propose a vulnerability approach to categorising class. According to their perspective, non-poor households that faced considerable risk could plausibly still slide into poverty and were thus not yet middle class. This distinction between vulnerability and security aligns with other important divides, including the separation between financial independence and reliance and a short and longer time horizon, which in turn has associations with savings behaviour and human capital investment decisions.

The vulnerability approach requires panel data to estimate the likelihood of falling into poverty based on a range of variables, including service delivery indicators such as access to running water as well as socio-economic characteristics of the head of the household such as educational attainment, employment status, gender and age. The middle class is defined as non-poor households who have a low probability of falling into poverty but who are still below an affluence threshold. The integration of multiple dimensions of deprivation is a key strength of this approach.

We apply the vulnerability approach by running probit regressions to estimate the likelihood of remaining in poverty or falling into poverty in the second wave of the NIDS 2010, based on the household's 2008 characteristics including the unemployment status of the household head, age of the household head (and its square), education of the household head (and its square), whether the household head is black, whether the household head is female, household size and an asset index.[8]

---

[8]The asset index is estimated using the multiple correspondence analysis command in Stata, based on variables relating to various aspects of the household's living conditions including access to water and sanitation and the energy sources used for lighting, heating and cooking, access to credit, home ownership and a long list of household possessions. The list of household possessions include car ownership and owning a television, microwave, radio, satellite, VCR, computer, camera, electric stove, gas stove, microwave, fridge, washing machine, sewing machine, knitting machine, lounge suite, motorcycle, plough, tractor, wheelbarrow and grinding mill.

Following Argent et al. (2009), we use per-capita income of R502 as the poverty line. Households are defined as lower class if the probability of becoming non-poor in 2010 is below 10%, vulnerable if the probability is between 10 and 50%, middle class if the probability is between 50 and 90%, and upper class if the probability is above 90%.

We find that 23% of households are very likely to be poor in both 2008 and 2010 and therefore considered to be chronically poor and to belong to the lower class. Table 2 also shows the strong association between race and class when applying the vulnerability approach. Almost 30% of black South Africans are categorised as lower class, a further 46% belong to the vulnerable class, 24% to the middle class and only 3% are categorised as upper class. In contrast, there are virtually no whites amongst the lower classes and the vulnerable, but 23% of white South Africans are categorised as middle class and 77% as upper class.

4.3 Income

Income is often used to estimate class because it provides a measure of an individual's economic power and is viewed as a correlate of social status because it determines a household's buying power and reflects market value and negotiating power in the labour market. Because income is a continuous variable, classes are defined using cut-off values of income. In economics, the middle class is often defined as a residual category, distinct from the lower classes (or the poor) and the upper class (or the affluent); two cut-off points are therefore required. Such points are usually defined based on percentiles, median or mean values or, alternatively, absolute thresholds.

Class can be defined using fixed percentages of the income distribution. For instance, Easterly (2001) defines the bottom 20% as lower class and the top 20% as upper class, thus categorising the remaining 60% as the middle class. One can also define poverty and affluence lines respectively as a share and multiple of a mean or median income. Birdsall et al. (2000) define the middle class as those with income between 0.75 and 1.25 of median per-capita income. Alternatively, we can use a poverty line set a specific level of per-capita income as the cut-off point to distinguish the middle class from the poor and, similarly, a line of affluence to distinguish the middle class from the affluent. This approach is sometimes preferred over the relative measures reliant on percentiles and medians because it allows comparisons with other countries.

However, all of these approaches remain vulnerable to the criticism that the imposed thresholds are arbitrary; we therefore opt for a polarisation method developed by Esteban et al. (1999) to find homogeneous social clusters using the underlying concepts of identification and alienation. The authors argue that traditional approaches utilising arbitrary cut-off points often result in groups without any internal cohesion and in which there is considerable variation in characteristics. This is likely to be a risk when applying arbitrary thresholds to South Africa's fragmented and unequal social landscape where there is little evidence of a large cohesive core. Using the patterns in the data to identify clusters and to make decisions about natural breaks and cut-off points, the polarisation method is more responsive to the anomalies and atypical features of South Africa's income distribution. Assuming that income patterns capture and reflect important differences in lifestyle and life chances that will correlate with patterns of socialisation and social identification, such an approach can offer an empirically grounded and defensible avenue to identifying a reasonably homogeneous and cohesive middle class within an unequal and divided society.

Table 3: Characteristics of classes when categorised using the income polarisation approach, 2008

Class	Share of total population (%)	Share of black population (%)	Share of white population (%)	Black share of each class (%)	Mean income (Rands)	Mean age	Mean educational attainment (years)
Lower class	60.6	70.7	4.7	92.2	376	37.7	7.6
Middle class	28.5	25.2	31.7	69.8	1 763	38.4	9.6
Upper class	10.9	4.1	63.7	29.4	9 574	43.1	12.5

Source: NIDS 2008. The classification of classes is based on the polarisation approach as described in Esteban et al. (1999). The mean age and education attainment relates to the household member with the highest occupational skill level.

As expected, Table 3 shows large gaps in the mean income per capita of lower-class, middle-class and upper-class households (R376, R1763 and R9573 respectively). The mean educational attainment of the highest skilled individual in the household varies according to class: it is 7.6 years for the lower classes, 9.6 years for the middle class and 12.5 years for the upper class. Both the lower and the middle classes are predominantly black (92% and 70% respectively), but the black share of the upper class is only 29%. One in four blacks is middle class.

4.4 Self-identification or subjective social class

Perceptions of individual ranking and social standing have long interested sociologists. Theorists such as Marx (Marx & Engels, 1968:37; Marx, 1972) and Durkheim (1933) assumed that individuals knew their position in society and that there was an alignment between objective and subjective social status. In contrast, the reference-group hypothesis raises the possibility that there may be a divergence between objective and subjective social class. It acknowledges that individuals' perceptions of their place in the social hierarchy are largely formed by their circle of close acquaintances (Stouffer et al., 1949). The crux of the argument is that the homogeneity of reference groups – the assortative tendency to surround oneself with friends of similar education, occupation and income – fundamentally distorts the subjective sample from which one generalises to the wider society and from which one develops perceptions of one's subjective location. Taken together, these lead to images of society with few at the top, the great majority in the middle and few at the bottom. In this view, perceptions of the shape of the social stratification system and of one's place in it are only loosely linked to objective circumstances, since objective conditions are filtered through the lens of reference groups.

We examine these theories by investigating self-identified social position using the NIDS data. All adult respondents were asked to imagine a six-step ladder with the poorest people in South Africa on the bottom or first step and the richest people standing on the top or sixth step. Respondents were then asked on which rung of the income ladder they saw their household. While the terms poor and rich would conventionally be interpreted in a narrow sense as referring to income, there is room for broader interpretations, especially given the use of the ladder metaphor.

Table 4: Characteristics of classes when categorised using self-reported assessment, 2008

Class	Share of total population (%)	Share of black population (%)	Share of white population (%)	Black share of each class (%)	Mean income (Rands)	Mean age	Mean educational attainment (years)
Lower class	33.9	37.5	13.3	87.7	771	38.4	7.6
Middle class	58.6	56.3	74.2	75.9	1 922	38.7	10.0
Upper class	7.5	6.2	12.5	65.2	4 434	38.6	11.3

Source: NIDS 2008. Households identifying themselves as standing on ladder rungs one and two were classified as lower class, those on ladder rungs three and four were classified as middle class and those on the top two ladder rungs were classified as upper class. The highest identified ladder rung was used to classify the household. The mean age and education attainment relates to the household member with the highest occupational skill level.

According to Table 4, perceptions of socio-economic position appear to be related to education level and income, but there is no association with age. There appears to be a tendency or desire to occupy a middle position rather than an extreme or outlying position in the social distribution. Assuming that objectively each rung of the ladder represents an equal proportion of the population, we can identify overrepresented categories. The analysis shows that 34% of the population place themselves on the bottom two rungs (33% based on equal share assumption), but a much higher than proportional 59% of the population place themselves on rungs three and four and a very low proportion (8%) see themselves as standing on one of the top two of society's ladder rungs. High social status individuals tend to underestimate their position in society.

Could this be explained by the reference-group hypothesis? In a country marked by deep and overlapping cleavages between white and black, affluent and poor, highly educated and low skilled, suburbs and townships, the reference-group hypothesis would lead a greater share of South Africans to identify themselves as standing on the middle rung of society's ladder when they are in fact at the top or the bottom. However, if race continues to serve as a good approximation for these overlapping social divides fostered by apartheid, then it is important to consider the distribution of self-identified social rank separately for black and white South Africans. For the black population, a slightly lower than proportional share views themselves as standing on the middle two rungs of the ladder (56%) and a slightly higher proportion on the lower two ladder rungs (38%) or the top two rungs (6%). Similarly, for the white population a much smaller proportion places themselves at the bottom of the distribution (13%), while 74% see themselves on the middle two ladder rungs and only 13% think that they occupy the top two rungs of society. This shows that both black and white South Africans of high social standing tend to underestimate their position in society. This is in line with Phadi & Ceruti's (2011) research, which showed that residents of Soweto preferred to see themselves as positioned in the middle of the social distribution, buffered from both sides by a more privileged and less privileged group.

While this analysis from the NIDS provides an interesting perspective on perceived social standing, it does not engage directly with the literature on self-assessed or

Table 5: Self-reported social class in South Africa, 1995, 2001, 2006 and 2013

Social class	1995 (%)	2001 (%)	2006 (%)	2013 (%)
Lower class	42.5	26.4	43.9	44.9
Working class	24.1	30.8	18.9	24.9
Lower middle class	16.6	21.0	19.5	17.3
Upper middle class	15.4	18.5	15.2	11.5
Upper class	1.4	3.2	2.4	1.4

Source: World values survey 1995, 2001, 2006 and 2013.

subjective class. Consequently, we also consider self-identified class based on the WVS of 1995, 2001, 2006 and 2013. Table 5 shows that the majority of South Africans consider themselves to be lower or working class. Those who identify themselves as middle class (including both the lower or upper middle class categories) grew from 32% in 1995 to 40% in 2001 before dropping to 35% in 2006 and 29% in 2013. A small minority of respondents consider themselves to be upper class in contemporary South Africa.

According to the 2006 WVS, those who identify themselves as lower class appears to be a distinct group with lower education and income levels. This group is also almost entirely black (97%). The distinction between those who identify themselves as working class or lower middle class appears to be more subtle, with similar age profiles, income levels, educational attainment and racial shares. The lines between upper middle class and upper class are also blurred with similar educational attainment, but with a slightly lower income level, lower age and higher black share for the upper class.

This work confirms earlier results by Seekings (2007) for a small sample of Cape Town residents showing that neighbourhood characteristics, race and occupation can only explain a modest share of the overall variation in subjective self-reported class. He concludes that 'the relationship between subjective and objective class was not tidy' (2007:30).

4.5 Comparing cross-section estimates with rival approaches

Viewing the vulnerability, occupation and income approaches as approximations of objective social status, we examined the misalignment between subjective and objective social status in more depth. When using the vulnerability approach to define classes, white South Africans are absent from the lower classes and there is only a negligible share of blacks who are classified as upper class. The individual-level data on self-reported social standing in the NIDS showed that there was a greater tendency to underestimate social standing amongst blacks (controlling for occupation, household income and education via regression analysis). This is in line with the qualitative work of Khunou (2012) and Krige (2012) suggesting that some young, educated and affluent black South Africans describe their membership of the middle class as tentative and conditional. The reluctance to use the middle-class label could be attributed to the term's strong historical association with being white. Alternatively, this evidence of greater vulnerability may link to Nieftagodien & Van der Berg's (2007) conclusion that the black middle class has a substantial asset deficit compared with other members of the middle class (Burger et al., 2014).

We use the 2006 WVS to analyse the correlates of self-identified class. Using regression analysis to examine the relationship of self-identified class and a list of characteristics that theory suggests should be linked to class (including a living standards index, education and occupation), it appears that the traditional, so-called 'objective' dimensions of class can explain only a small share of how individuals self-identify into classes. Our analysis suggests that a multi-dimensional living standards index has the most powerful association with self-identified class and explains about 22% of the observed variation in self-identified class. Years of education is also significant, but only explains 8% of the observed variation in self-reported class. Occupations and the manual-cognitive content of work both have a significant association with self-reported class, but do not explain much of the variation.

Similar to the findings of Lopez-Calva et al. (2011b), we find little empirical basis for the widely held belief that certain values and attitudes are robustly and significantly associated with the middle class. This also resonates with the findings of Mattes (forthcoming), which show virtually no significant relationship between middle-class membership[9] and patriotic or democratic values. Independence, financial prudence and imagination were significantly associated with the likelihood of respondents classifying themselves as middle class, but these values did not contribute much in terms of explanatory power nor was there a significant association between these values and the likelihood of identifying as middle class amongst the black sample. The only noteworthy exception to this rule is our finding that black South Africans who regard religion as important are significantly more likely to describe themselves as middle class.

5. Comparing trends in the middle class for rival approaches

We compare trends in the growth of the black middle class applying our four approaches to three waves of the NIDS and the 1993 PSLSD. We also examine self-reported class according to the WVS in 1995, 2006 and 2013.

Unfortunately, the 1993 PSLSD has a large number of missing values for occupation and we were thus forced to exclude this occupation class estimate for this survey. To ensure the comparability of estimates of the middle class, we had to implement a simplified version of the vulnerability model[10] using a linear probability model instead of a probit. We implement the income approach by adjusting the cut-off points estimated in the 2008 NIDS for inflation.

Based on occupation there has been only a modest increase in South Africa's black middle class share between 2008 and 2012; it increased slightly from 35% in 2008 to 37% in 2012. Because we do not have estimates for 1993 for occupational class and because the estimated trend is very flat for the rest of the years, we do not include these numbers in the figure.

Figure 1 shows a dramatic increase in the proportion of black South Africans classified as middle class based on both the income and the vulnerability approach, but with a more

---

[9]Mattes defines the middle class as individuals 'whose salaries and social milieu enable them to live middle class lifestyles' (forthcoming) and who are never without the basic necessities (e.g. water, food, healthcare).

[10]Here we use a more parsimonious asset index (cf. model in Section 4.2) because we are restricted to assets that occur both in the 1993 PSLSD and the NIDS panel. We implement asset terciles using access to clean water, adequate sanitation, electricity, fridge, television or radio and stove.

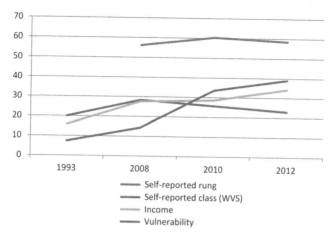

Figure 1: Comparing trends in rival approaches for 1993, 2008, 2010 and 2012 (or closest years)
Notes: NIDS 2008, 2010 and 2012; and SA World Values Survey for 1995, 2006 and 2013. We report interpolated WVS values for comparison with the NIDS 2010 estimates.

pronounced rise in the black middle class according to the vulnerability approach than when using an income measure. The steep rise in the black middle class shown by the vulnerability approach (from 7.4% in 1993 to 39.0% in 2012) may be partly due to improvements in the circumstances and characteristics associated with greater financial security such as educational attainment, access to clean water and electricity and ownership of stoves and fridges. When class is measured based on per-capita household income cut-off points (as described in Section 4.3), the increase in the black middle class is less pronounced than in the case of the vulnerability approach, but the rise remains sharp from 15.9% in 1993 to 34.2% in 2012.

It is illuminating to compare these more 'objective' indicators with trends in how individuals described their own social position. As reported before, we find that individuals overwhelmingly tend to place themselves in the middle of the income distribution, with about 60% of respondents identifying themselves to be on the middle two rungs of the income distribution. Although one should be cautious to compare the NIDS estimates with those from the WVS, the magnitude of the gap is wide enough that it is not controversial to observe that many South Africans who think of themselves as being in the middle of the income distribution do not describe themselves as middle class, indicating that this is a more loaded term and is not simply seen as referring to those who see themselves occupying the middle of the income distribution.

Because the self-reported rung question is based on a relative classification of social position and income, one would not expect it to be responsive to absolute improvements in income or security. However, one may expect some correspondence between the self-reported class and our 'objective' class indices such as the vulnerability approach and the income measures. It is therefore interesting to see that the sharp observed rise in the black middle class shown by these 'objective' indicators is not mirrored in the proportion of black South Africans who classify themselves as middle class over this period. This is in line with previous studies showing that individual conceptions of what class membership means are slow to change, but more significantly that there is a sluggish response of identity to changes in the individual's 'class markers'. This could be related to an inherent reluctance to reclassify oneself, especially in a context where class has a strong link to social identity or may confirm Inglehart & Welzel's (2005) ideas around the socialisation aspect of class and the dominant role of childhood experiences and circumstances.

6. Conclusion

Our analysis shows considerable variation in who is classified as middle class based on the different definitions that we employ. While there is some overlap, there are also substantial differences in who is identified as the middle class based on whether we implement the occupational, income, vulnerability or subjective social status approach to discern class membership. Reported income, occupation and vulnerability measures appear to have a weak relationship with self-identified class. This is in line with the work of Schlemmer (2005), who argued that the South African middle class lacks cohesion and that skilled and affluent individuals are reluctant to self-identify as middle class. Qualitative work by Khunou (2012) and Krige (2012) suggests that well-educated, skilled black South Africans appear uncomfortable with the 'middle class' tag and are hesitant to describe themselves in this way – seemingly at least partly because of the category's strong historical association with being white and consequent perceived tensions between their racial and cultural identity and persistent ideas around what being 'middle class' represents.

The country's apartheid past, and the post-apartheid state's responses to this, obviously strongly influenced both the nature of the middle class in contemporary South Africa and individual behaviour of members of this class. The perceived vulnerability of sections of the black middle class has much to do with the fact that they lack the assets that allow them to fully adopt a middle-class lifestyle, but also with the tenuousness of their middle-class status. Stronger economic growth would perhaps have engendered greater confidence in their ability to maintain their status, but modest economic growth over most of the post-transition period and economic shocks may have undermined greater security.

State policies greatly influenced the growth and the nature of the black middle class. Economic empowerment created a small but visible group of black capitalists. More important in numerical terms was upward social mobility in both the public and private sectors, with affirmative action policies lending support. The beneficiaries were those best placed to enter the higher rungs of the labour market. The immediate post-apartheid period and some catch-up allowed a large expansion of this group, but the weak performance of both the education system and the economy does not augur well for similarly rapid expansion in future. This creates an interesting political dynamic, with younger cohorts expecting more well-remunerated jobs than the education system and labour market may be able to provide.

Social identity is a crucial factor in the pursuit of greater social cohesion and political stability. If the racial gap in earnings and education is decreasing, but racial differences continue to loom large in the minds of South Africans, the growth in the black share of middle-class professions or the black share of income may not translate to a more integrated and less polarised social and political landscape. This may also be an important finding for the wider literature, demonstrating that it is vital to move beyond the surface and explore how class categories are perceived and used. Research on class should consider not only objective social status, but also subjective social status – especially in light of its potential role in mediating key economic and political benefits associated with the middle class.

Funding

This work was supported by the Vice-rector of Stellenbosch University's discretionary fund.

References

Argent, J, Finn, A, Leibbrant, M & Woolard, I, 2009. Poverty: Analysis of the NIDS Wave 1 dataset. NIDS Discussion Paper No. 13, South African Labour and Development Research Unit (SALDRU), School of Economics, University of Cape Town.

Beckett, A, 2010. Is the British middle class an endangered species? The Guardian, 24 July, p. 28. http://www.guardian.co.uk/uk/2010/jul/24/middle-class-in-decline-society Accessed 2 July 2013.

Birdsall, N, 2010. The (indispensable) middle class in developing countries; Or, the rich and the rest, not the poor and the rest. Working Paper 207, Center for Global Development, Washington, DC, USA.

Birdsall, N, Graham, C & Pettinato, S, 2000. Stuck in the tunnel: Is globalization muddling the middle class. Working Paper No 14, Center on Social and Economic Dynamics, Washington, DC, USA.

Burger, R, Louw, M, Pegado, B & Van Der Berg, S, 2014. Understanding the consumption patterns of the established and emerging South African black middle class. Development Southern Africa 32(1). doi:10.1080/0376835X.2014.976855

Ceruti, C, 2013. A proletarian township: Work, home and class. In Alexander, P, Ceruti, P, Motseke, K, Phadi, M & Wale, K, Class in Soweto. University of KwaZulu-Natal Press, Scottsville.

Crankshaw, O, 1997. Race, Class and the Changing Division of Labour Under Apartheid. LSE, London.

Durkheim, E, 1933. The Division of Labour in Society. Free Press, Glencoe.

Easterly, W, 2001. The middle class consensus and economic development. Journal of Economic Growth 6, 317–35.

Esteban, J & Ray, D, 1994. On the measurement of polarization. Econometrica 62(4), 819–52.

Esteban, J, Gradín, C & Ray, D, 1999. Extensions of the measure of Polarization, with an application to the income distribution of five OECD countries. Working Paper No. 218, Maxwell School of Citizenship and Public Affairs, Syracuse University, New York.

Evans, G & Mills, C, 1999. Are there classes in post-communist societies? A new approach to identifying class structure. Sociology 33(1), 23–46.

Glassman, RM, Swatos, WH & Kivisko, P, 1993. The Noble Character and Flaws of the Middle Class. Greenwood Publishing Group, Westport, CT.

Goldthorpe, JH, 1987. Social Mobility and Class Structure in Modern Britain. Clarendon Press, Oxford.

Goldthorpe, JH & Erikson, R, 1993. The Constant Flux. A Study of Class Mobility in Industrial Societies. Clarendon Press, Oxford.

Goldthorpe, JH & McKnight, A, 2004. The economic basis of social class. CASE Paper 80, Centre for Analysis of Social Exclusion, London School of Economics and Political Science, London.

Inglehart, R & Welzel, C, 2005. Modernization, Cultural Change and Democracy: The Human Development Sequence. Cambridge University Press, Cambridge.

Khunou, G, 2014. What middle class? The shifting and dynamic nature of class position. Development Southern Africa 32(1).

Krige, D, 2014. 'Growing up' and 'moving up': Metaphors that legitimise upward social mobility in Soweto. Development Southern Africa 32(1).

Lopez-Calva, LF & Ortiz-Juarez, E, 2011a. A vulnerability approach to the definition of the middle class. Policy Research Working Paper Series 5902, The World Bank, Washington, DC, USA.

Lopez-Calva, LF, Rigolini, J & Torche, F, 2011b. Is there such thing as middle class values? Class differences, values and political orientations in Latin America. Policy Research Working Paper 5874, The World Bank, Washington, DC.

Mattes, R. Forthcoming. South Africa's emerging black middle class? Journal of International Development.

Marx, K. 1867. Capital: A Critique of Political Economy. Charles H. Kerr and Co., Chicago.

Marx, K, 1972. Economic and philosophic manuscripts of 1844: Selections. In Tucker, RC (Ed.), The Marx-Engels Reader. W.W. Norton, New York.

Marx, K & Engels, FC, 1968. The Communist Manifesto. International Publishers, New York.

Nieftagodien, S & Van der Berg, S, 2007. Consumption patterns and the black middle class: The role of assets. Stellenbosch Economic Working Paper 02/07, Bureau for Economic Research & Department of Economics, University of Stellenbosch.

Pakulski, J, 2005. Foundations of a post-class analysis. In Wright, EO, Conclusion: If Class is the Answer What is the Question. Approaches to Class Analysis. Cambridge University Press, Cambridge.

Phadi, M & Ceruti, C, 2011. Multiple meanings of the middle class in Soweto, South Africa. African Sociology Review 15(1), 88–108.

Rivero, C, Du Toit, P & Kotze, H, 2003. Tracking the development of the middle class in democratic South Africa. Politeia 22(3), 6–29.

Schlemmer, L, 2005. Lost in transformation? South Africa's emerging middle class. CDE Focus Occasional Paper No 8, Centre for Development and Enterprise, Johannesburg, South Africa.

Seekings, J, 2003. Do South Africa's unemployed constitute an underclass? Working Paper No. 32, Centre for Social Science Research, Cape Town, South Africa.

Seekings, J, 2007. Perceptions of class and income in post-apartheid Cape Town. CSSR Working Paper No. 198, Centre for Social Science Research, University of Cape Town.

Seekings, J, 2009. The rise and fall of the Weberian analysis of class in South Africa between 1949 and the early 1970s. Journal of Southern African Studies 35(4), 865–81.

Seekings, J & Nattrass, N, 2005. Race, Class and Inequality in South Africa. Yale University Press, New Haven, CT.

StatsSA (Statistics South Africa), 2009. Profiling South Africa middle class households, 1998–2006. Statistics South Africa Report 03-03-01, Statistics South Africa, Pretoria.

Stouffer, SA, Suchman, EA, de Vinney, LC, Star, SA & Williams, JR, 1949. The American Soldier: Adjustment during Army Life. Princeton University Press, Princeton, NJ.

Van der Berg, S. 2010. The demographic and spatial distribution of inequality. In Bernstein, A (Ed.), Poverty and Inequality: Facts, Trends and Hard Choices. Centre for Development and Enterprise Round Table Paper Number 15, Centre for Development and Enterprise, Johannesburg.

Visagie, J & Posel, D, 2013. A reconsideration of what and who is middle class in South Africa. Development Southern Africa 30(2), 149–67.

Weber, M, 1961. Essays in Sociology. Routledge and Kegan Paul, London.

Weber, M, 1968. Economy and society: An outline of Interpretive Sociology. Bedminster Press, New York.

Wright, EO, 1979. Class, Crisis and the State. Verso, London.

Wright, EO, 1980. Class and occupation. Theory and Society 9, 177–214.

Wright, EO, 1989. The Debate on Classes. Verso, London.

Understanding consumption patterns of the established and emerging South African black middle class

Ronelle Burger[1], Megan Louw[2], Brigitte Barbara Isabel de Oliveira Pegado[3] & Servaas van der Berg[4]

Existing empirical research on consumption patterns of the South African black middle class leans either on the theory of conspicuous consumption or culture-specific utility functions. This paper departs from treatment of the black middle class as a homogeneous group. By differentiating between a securely established group, with characteristics and consumption patterns similar to the white middle class, and an emerging group, often with weaker productive characteristics, the paper formally introduces economic vulnerability as a driver of consumption patterns. Households new to the middle class or uncertain of continued class membership are viewed as vulnerable. Consumption patterns of the emerging black middle class are observed to diverge substantially from the other groups, in terms of greater signalling of social status via visible consumption and preoccupation with reducing an historical asset deficit. We expect many of its members to join the established classes over time, converging to a new 'middle class mean'.

1. Introduction

Much has been said about the levels of income inequality in South Africa, which do indeed remain amongst the worst in the world. Thabo Mbeki, soon thereafter president, described it eloquently back in 1993 when he first referred to South Africa as a country of two nations: one wealthy and historically white; and the other completely excluded from the economic mainstream, impoverished and black (Mbeki, 1998). After 20 years of democracy a large proportion of South African remains excluded from the labour market and the mainstream economy. South Africa's unemployment rate and Gini coefficient have remained stubbornly high. Increases in the frequency and violence of protest marches calling for improved service delivery and better economic conditions may be evidence that frustrations amongst the marginalised are escalating (Alexander, 2010; Office of the President, 2013).

[1]Associate Professor, Department of Economics, University of Stellenbosch, Schumann Building, Bosman Street, Stellenbosch 7600, South Africa.
[2]Researcher, Department of Economics, University of Stellenbosch, Schumann Building, Bosman Street, Stellenbosch 7600, South Africa.
[3]Researcher, Department of Economics, University of Stellenbosch, Schumann Building, Bosman Street, Stellenbosch 7600, South Africa.
[4]Professor of Economics and South African Research Chair in the Economics of Social Policy, ReSEP, Department of Economics, University of Stellenbosch, Private Bag X1, Matieland, Stellenbosch 7602, South Africa.

The enduring problems with unemployment and the poor quality of service delivery sometimes mask the post-apartheid successes. This period has also seen a considerable expansion in service delivery coverage and a rapid increase in black affluence. In an article in this issue, Visagie shows that an additional 3.1 million black people were added to the middle class between 1993 and 2008 (see also Mamabolo, 2013). This shift is significant not only for markets and for economic growth, but in a broader societal context can be interpreted as an indication that South African society has become more open and dynamic. Furthermore, a growing middle class is considered socially beneficial by Easterly (2001) and others due to its association with desirable outcomes such as more prudent policy, more education and improved political stability and democracy.

This paper attempts to contribute to a better understanding of the position and the impact of the rising black middle class on the South African economy by investigating the consumption preferences and choices of this growing consumer group. What requires explanation is why consumer patterns of the black middle class differ from those of their white peers. Following Bourdieu's (1984) view that dominant tastes are a reflection of social power of different classes as reflected in their cultural capital, one would expect a convergence in the patterns of consumption of individuals entering the middle class to that of the dominant white middle class.

Much of the initial research on the purchasing decisions of the black middle class group was conducted by consumer market researchers eager to define and describe such spending habits in terms of new consumer categories and emphasising differences in underlying utility functions ('tastes') between race groups that they often appear to regard as immutable. However, the consumer market focus on preferences and tastes has been criticised for perpetuating narrow stereotypes of the black middle class as conspicuous consumers with a taste for expensive cars, designer labels and large houses and a reputation as poor creditors. There have been only a handful of studies looking for more fundamental explanations for differences in spending patterns of the emergent black middle class in South Africa, notably Nieftagodien & Van der Berg (2007) and Kaus (2013).

The aim of this paper is to provide more encompassing hypotheses for why more vulnerable members of the middle class may exhibit different spending priorities. Significantly, these rival explanations find the rationale for the differential spending patterns of the emergent middle class in their vulnerable circumstances and their asset deficit, rather than in their unique intrinsic preferences. Such a conceptual shift may be subtle, but has important implications for anticipated trends. If differential spending patterns are attributable to intrinsic differences, then such gaps will remain, whereas if they are due to the vulnerable new entrant position of the emergent black middle class, then such gaps will fade and dissipate over time.

A first applicable theory relates to conspicuous consumption, as put forward by Veblen late in the nineteenth century. Under this theory, individuals derive utility from the social status linked to visible consumption of certain goods. Conspicuous consumption is thus consumption that is intended to be visible. This is in effect signalling wealth, which is generally unobserved. It is relative in the sense that an individual's status is socially contingent; that is, relative to that of other individuals within the reference group. In South Africa, race can be used to define the reference group, since race has played a significant role in forming cultural and economic identities. Kaus (2013) finds variance in conspicuous consumption amongst groups, specifically 35 to 50% greater expenditure on visible consumption amongst coloured and black households relative to whites, linking this to a signalling model of social status.

A second applicable theory is that expenditure patterns within less affluent groups are driven by historical asset deficits. Black, coloured and Indian households are playing 'catch up' with the household asset levels typical of more established white middle class households.

This paper builds on previous studies demonstrating an economic perspective to show that expenditure patterns are primarily driven by rational and dynamic socio-economic factors and that one need not resort to explaining differences in underlying utility functions (tastes) or any other deterministic racial grouping attributes. These include status seeking in a broader sense as well as reducing the asset deficit. Together, these factors explain a significant portion of the observed variation in consumption expenditure across black and white members of the middle class.

Following a literature review, definitional issues are discussed and the methodology applied in this paper is presented. Results of descriptive analysis and modelling on the Income and Expenditure Survey 2010/11 (IES) suggest that greater conspicuous consumption observed amongst the black group reduces when taking into account new membership or uncertainty regarding future membership of the middle class. As the emerging black middle class establishes itself and members continue to transition into the established group, it is likely that substantial convergence to a new South African 'middle class mean' will take place.

2. Understanding the term 'conspicuous consumption'

Much of the initial work on the emergent black middle class viewed this group primarily as an attractive consumer market, depicting its members as highly ambitious and aspirational in their spending patterns. The foundations for this belief are found in the group's investment in education, its robust expenditure underpinned by the recent growth in credit extension, and rising incomes associated with black economic empowerment policy, upward social mobility and economic growth. Krige (2009, 2011) notes that commentators and researchers have criticised the work on so-called 'Black Diamonds' as propagating cultural stereotypes. Members of the black middle class argue that the black middle class is incorrectly painted as greedy and consumerist, resulting in what is seen as a simplistic, patronising and inaccurate representation of reality. Such a stylised caricature of black middle-class consumers suggests instead that membership of a specific subgroup is the most important explanation for observed consumer patterns, obscuring the role of rational economic drivers of consumption expenditure.

Veblen's economic theory of conspicuous consumption proposes that individuals gain social status by signalling their wealth to their reference group by engaging in conspicuous leisure or conspicuous consumption, the latter defined as visible consumption of certain goods that are associated with social status (Trigg, 2001; Kaus, 2013). Veblen (1899) argued that individuals demonstrate their wealth and so gain in social status by showing that they can afford to waste time, effort and money. Veblen, however, also noted that amongst the more established members of the upper class the need to signal wealth may be diminished, partly because they can signal their wealth through a set of distinct habits and tastes which they acquired tacitly through their social upbringing.

This insight is central to the more recent work of Bourdieu (1984) on how tastes and preferences can signal and entrench class. He argues that taste and preferences are class markers that distinguish and legitimise privilege. Each class therefore aspires to mimic the tastes and consumption patterns of those above it. According to such a perspective, the established upper classes would be less likely to exhibit conspicuous consumption while it would be important for the middle class to distinguish themselves from the working class (Trigg, 2001).

Bourdieu (1984) believes that displays of consumption need not be crude and deliberate, but can often work through a tacit code acquired through socialisation and motivated by socialised norms and tendencies that guide behaviour and thought (Lamont & Lareau, 1988). Linked to this theory is the notion of cultural capital, which provides an avenue through which groups can express domination within the social hierarchy through the claim of possessing 'good taste'. Bourdieu's (1984) conceptualisation of the complex social constellation that implicitly governs and guides the choices of the members of social groups allows for a more intricate, but also more fluid, view of class structure and expenditure patterns.

3. Previous empirical investigations of the relationship between consumption and class

The work of Veblen and more recently Bourdieu has inspired a large body of empirical research examining the relationship between conspicuous consumption and class. Charles et al. (2009) and Kaus (2013) postulate that conspicuous consumption is socially contingent in the sense that an individual's visible expenditure is influenced by the characteristics of the reference group and his/her position within the income distribution of this reference group.

These authors examine whether conspicuous consumption will increase as the mean income of the reference group decreases. The intuition is that an individual from a relatively poor reference group who aspires to achieve social status of higher groups will have an incentive to engage in additional signalling, given the general perception of low status associated with this group.

They also investigate whether conspicuous consumption will increase as income inequality within the reference group increases. The intuition here is that additional signalling may be required to demonstrate positioning near the top of the reference group income distribution. Lastly, they explore whether conspicuous consumption increases with the permanent income of the relevant household.

Charles et al. (2009) search for inter-racial evidence of conspicuous consumption in US data, using nationally representative household surveys and panels. Defining conspicuous consumption as expenditure on items such as cars, clothing and jewellery, they find that black and Hispanic individuals devote substantially larger shares to such conspicuous consumption, consistent with a model of status seeking. The trade-off associated with higher visible consumption is lower consumption of all other categories of current consumption – notably education and health – as well as future consumption. This implies a significant and intertemporal cost to conspicuous consumption in terms of other consumption foregone.

The authors examine whether this conspicuous consumption is associated with membership of particular reference (race and regional) groups. They find that visible consumption increases with income and the dispersion (inequality) of reference group income, while it decreases with reference group income (Charles et al., 2009). Since a socially contingent model of conspicuous consumption explains much of the observed variation in conspicuous consumption, the authors therefore do not give much weight to deterministic factors such as racial differences in tastes.

Kaus (2013) follows a similar methodology but focuses instead on South Africa using the Income and Expenditure Surveys of 1995, 2000 and 2005. He finds that coloured and black households spend 35 to 50% more on visible goods than do comparable white households.[5] When exploring the status-seeking model as an explanation for this consumption, Kaus also finds that visible consumption is higher when reference group income is lower. However, the results are not replicated within all race groups, and he therefore does not rule out that there are differences in underlying tastes and preferences. Trade-offs occur through lower spending on health, housing, entertainment and communications amongst black and coloured households.

Whilst these studies link conspicuous consumptions to specific income characteristics of social groups, they do not take into account the potential impacts of small movements across a class threshold on expenditure patterns. Since income-based approaches to the middle class assume enjoyment of an aspirational lifestyle associated with incomes above a certain threshold, proximity to this point should be an important driver of expenditure patterns within the middle class. An innovative new study by Lopez-Calva & Ortiz-Juarez (2011) examines vulnerability to poverty in the context of a rising middle class in Central and Latin America, observing that a group of households located between the middle class and the poor in the income distribution remains vulnerable to falling back into poverty. This may be explained in terms of structural characteristics of these households. Applying the Weberian view of class, these individuals are grouped into classes according to common economic 'life chances' that influence their market income opportunities. The middle class is defined as a group benefiting from a broad skills base and substantial investment in education, offering it a significant chance of attaining economic prosperity over time.

The income-based approach to definitions of the middle class necessarily segments the population independently of productive characteristics such as educational attainment. Accordingly, there may be 'subgroups' within the affluence-based middle-class group that exhibit fundamentally different characteristics associated with social mobility, thus enjoying very different 'life chances'. In a similar vein, Goldthorpe & McKnight (2004) distinguish between groups of workers on the basis of economic security, stability and prospects (Lopez-Calva & Ortiz-Juarez 2011). This echoes Ravallion's (2010) work, which points out that the growing middle class in developing countries remains vulnerable, despite its newly acquired relative affluence. This vulnerability itself may be an important driver of expenditure patterns, should a household feel insecure due to recently joining the middle class or lacking sufficient productive resources to be confident of sustaining class membership.

A separate explanation for differing racially based expenditure patterns is advanced by Nieftagodien & Van der Berg (2007), framing the issue in terms of an 'asset deficit'.

---

[5]However, note that this effect is not observed for cars in the South African context (Kaus, 2013:20).

They argue that the ramifications of past economic racial segregation could explain currently observed differences in asset levels, with black and coloured households consequently at a disadvantage in terms of ownership of household assets. In white households, there is likely to be some intergenerational transfer of these goods, enabling new generations to allocate a greater share of their disposable income to other items. Their research shows that black middle-class households have asset deficits and are more likely than their white counterparts to purchase assets.

4. Data

This paper mainly uses the IES for analysis. This survey was conducted by Statistics South Africa during the period of September 2010 to August 2011. Data were recorded for a sample of 25 328 households across the country over a 12-month period. The IES is mainly conducted to provide statistical information on household consumption expenditure patterns for the calculation of the weights for the consumer price index. A combination of diary and recall methods was used to sample these households. Each household was presented with a questionnaire and a two-week diary. The diary acquisition approach may generate underlying measurement error due to the fact that respondents may become fatigued or forget to complete their diaries on a daily basis – thus they may not comprehensively record all actual expenditures over the two-week period (Visagie & Posel, 2013).

Although the primary aim of the IES is to provide information for the readjustment of the consumer price index basket of goods and services, a secondary aim is to provide additional information on socio-economic conditions. In this paper, additional information on the presence of assets is also used to create two indices to rank middle-class households in the context of socio-economic conditions.

5. Analysis of conspicuous consumption in the South African middle class

5.1 Defining the middle class

The analysis uses an income-based definition of the middle class to examine what consumption patterns reveal about the evolving social landscape and de-racialisation in post-apartheid South Africa. It investigates the distinct consumption patterns of households enjoying at least middle-class lifestyles, namely those that have enough income that they no longer need struggle with the necessities and the basics, but have sufficient income to make allocations towards discretionary expenditure. We consequently employ an income-based approach in applying this definition. Only a lower income cut-off value is applied, set at a level that yields a middle class comprising the most affluent 15% of the population, namely R53 217 per capita per annum in 2010/11 terms.

An upper income cut-off value is not applied, because such an incision leaves only a slither of very rich at the top. Additionally, because of the small number of observations in what would be the upper-income group within the black population, it is difficult to meaningfully analyse such a category over time or to decompose it by other attributes and qualities. While the distinction between the middle and upper classes may be of interest for understanding capital formation or the distributional effects of growth, the wealth of the upper classes cannot be captured and represented reliably and accurately without specialised wealth surveys.

5.2 Distinguishing the emergent and the established middle class

Based on asset levels, we then distinguish between two subgroups of the middle class, namely the emerging and established middle class. The purpose of this distinction is to allow for a more nuanced, dynamic analysis of spending patterns that implicitly takes into account the duration and degree of class inclusion. In describing the black middle class in 2005, Schlemmer (2005:5) refers to lagging security with respect to asset ownership, status and self-confidence within the group, still very new and small then.[6] It is a hypothesis of the current analysis that such insecurity remains a characteristic feature of the emerging component of the black middle class, associated with households that have recently joined the middle class or whose grasp on middle-class status is tenuous, for example, due to unstable forms of income. This builds on the approach taken by Schlemmer, which refers to a 'core middle class' comprising households containing highly skilled workers such as professionals and managers, as distinct from a 'lower middle class' made up of more modestly salaried clerical workers, teachers, nurses, and so on. Specifically he observed a notable lack of unity in class identity in this group with many individuals failing to identify themselves as belonging in the middle class.

The work by Nieftagodien & Van der Berg (2007) suggests that a high expenditure priority for new members to the middle class is the acquisition of assets typically associated with middle-class lifestyles, such as white goods and cars. By this rationale, established households that have a longer membership of the middle class have accumulated many of the goods traditionally found in white middle-class households, and should thus score higher on the asset index. Conversely, new entrants to the middle class will face an asset deficit.

To provide further evidence of the asset deficit amongst new entrants to the middle class, we turn to data from the 2012A version of the All Media and Products Survey on ownership and purchases of large household appliances. This survey shows that white households with a monthly income of at least R8 000 are six times more likely than their equally affluent black counterparts to own a tumble dryer, whereas the latter are five times more likely to have recently acquired one. Similarly, black households in this income category are only one-half as likely as white households to own a washing machine, yet responses regarding purchasing patterns indicate that they appear four times more likely to have recently bought a washing machine. There is also evidence that they are significantly more likely to purchase microwaves and refrigerators. This provides some evidence that relatively affluent black households experience an asset deficit compared with whites of similar income, but they are allocating more of their resources to eliminating this deficit.

To create a single indicator variable to collectively represent the assets of a household, we create an index from binary variables reflecting ownership of white goods and other household assets. These include televisions,[7] DVD players, refrigerators, stoves, microwaves, washing machines, motor vehicles, computers, cameras, telephones, satellite dishes, the Internet, furniture and ownership of a brick house. Of the three most common methods used to construct an asset index – namely factor analysis,

---

[6]Almost one-third of black middle-class survey respondents identified themselves as 'working class' (Schlemmer, 2005).
[7]We conducted sensitivity testing due to concerns about the inclusion of expenditure on television in the conspicuous consumption share and television ownership in the asset index, but excluding television from the asset index does not influence the results presented here in any meaningful way.

principal component analysis and multiple correspondence analysis (MCA) – MCA is selected because it imposes fewer restrictions on the survey matrix (it does not, for instance, assume a normal distribution of the underlying variables) and because it is suited to the use of categorical variables (Booysen et al., 2008). Using MCA as a data reduction technique, an asset index is constructed and black households within the middle class as defined here are then divided on the basis of this index into those with more assets (constituting a more established grouping within the middle class) and those with fewer assets, referred to as an emerging middle class.

5.3 Defining conspicuous consumption

Our analysis defines conspicuous consumption as expenditure on categories of goods and services that may be considered luxury items associated with affluent lifestyles. By its very nature, conspicuous consumption will be defined to include different items according to the consumer's social context since it is a socially contingent concept. In our paper focusing on South African consumers, both Veblen's theoretical framework for conspicuous consumption and the practical constraints arising from survey data quality and accuracy informed its measure. Conspicuous consumption was accordingly defined to include expenditure associated with clothing, footwear, restaurants, grooming products and services, watches, handbags, televisions and satellite dishes.[8] However, when interpreting this it is important to bear in mind that even the poorest households will allocate some money towards footwear, clothing and grooming. Only that part of expenditure on these items which exceeds functional levels can properly be regarded as an attempt to signal wealth. Therefore, this variable should be interpreted in a relative sense. Some of what is interpreted as conspicuous consumption could be necessary and functional investment in household assets and other goods.

5.4 Descriptive analysis

Firstly, a descriptive analysis is presented in Table 1, reflecting some interesting contrasts between the subgroups comprising the middle class.

As anticipated, black established and white middle-class groups look quite similar in terms of productive characteristics. Approximately one-half of households in these groups have at least a diploma, while more than 20% have a degree. Similarly, household heads are older and probably well established in their careers, reflected in high scores on the asset index. The most obvious difference is the much higher per-capita income level of white household heads, perhaps due to a combination of more experience (household heads are older), greater historical representation in financially rewarding occupations, more profitable social networks (greater social capital, in Bourdieu's terms) and more reliance on passive income streams. This last factor is

---

[8]One may argue that purchasing a car is a necessity in a country such as South Africa, which historically has offered individuals limited options for reliable and convenient public transport, particularly within urban areas. Therefore, only excessive expenditure on (luxury) cars can be regarded as conspicuous consumption. Given limited information on the nature and value of car purchases in the data and the fact that the large value of car purchases would result in it dominating conspicuous consumption spending patterns, car spending was not included in our analysis. This deviates from the work of Charles et al. (2009) and Kaus (2013), who included expenditure on cars as conspicuous consumption in the American and South African contexts respectively.

Table 1: Descriptive characteristics of the middle class

	Middle class	Black emerging middle class	Black established middle class	White middle class
Total population (total South African population = 50 375 543)	7 553 951	1 496 533	1 293 608	3 438 123
% of South African population	15.00	2.97	2.57	6.82
Household head has at least a degree (%)	19.48	9.09	27.77	22.52
Household head has at least a diploma (%)	43.29	28.45	56.49	48.57
Education of household head (years)	11.88	10.71	12.48	12.34
Age of household head	47.0	42.3	45.2	49.9
% households with female head	18.95	28.61	21.12	15.38
% household heads employed	80.9	88.7	84.6	73.8
% household heads self-employed	9.4	7.5	7.6	11.9
% rural	8.0	20.1	9.1	5.2
Average household size	3.55	3.02	4.19	3.38
% of households single members	8.8	24.5	2.7	6.2
Average per-capita income (Rands)	138 673	97 036	120 603	171 203
Average per-capita expenditure(Rands)	96 268	47 331	81 880	131 372
Average conspicuous consumption(Rands)	16 924	10 830	20 673	18 034
Mean value of asset index (mean = 0, standard deviation = 1)	0.210	−0.993	0.472	0.562
Conspicuous consumption including new car purchases (Rands)	31 483	16 001	39 318	34 508
Conspicuous consumption share (average across households) (%)	6.95	10.67	7.79	4.99

Source: Authors' own calculations from the IES.

borne out by the much lower employment rates of white household heads, which is not fully explained by slightly higher self-employment rates.

By contrast, emerging black middle-class households display substantially less entrenched prosperity. Household heads are younger and thus less experienced, and are much less likely to have tertiary education than their peers, suggesting lower human capital overall. Furthermore, emerging black households are more likely to be female-headed, single-person households, and to be located in rural areas. In the South African context, all of these characteristics are typically associated with lower incomes and greater vulnerability. It is thus unsurprising that scores on the asset index are much lower, in line with relatively low per-capita incomes.

It seems quite likely that a portion of the black emerging group – typically with young, well-educated household heads and located in urban areas – will transition into the black established group over time, as they gain experience in the workplace and are promoted into more lucrative positions. However, a significant portion of this group is structurally less advantaged and will thus remain marginal members of the middle-class group, their social mobility potential capped by lack of access to opportunities in urban labour markets, or by not having an income-earning spouse, for example.

Interestingly, and in line with the hypothesis, the black emerging group allocates more expenditure towards both conspicuous consumption items and assets than either of the other subgroups.

5.5 Regression analysis

We perform regression analysis to determine whether expenditure patterns are significantly different between groups once control variables are added. The analysis applies an augmented version of the permanent income hypothesis to model spending patterns. Formally, we model conspicuous consumption expenditure share as:

$$Y = \beta_0 + \beta_1{}^*r_i + \phi^*p_i + b^*a + \theta X_i + \eta_i$$

where r is a range of racial dummy variables, p represents the household's permanent income (instrumented here by per-capita household income), 'a' represents the score on the asset index, and X_i is a vector of controls.

Dummy variables indicate whether the head of a household is black, Indian or coloured (the reference group is white household heads). Other standard controls that may proxy for differences in prices, tastes and preferences include a dummy variable for rural location, the number of children in the household, the number of elderly people in the household and a dummy indicating whether the head of the household is female. Per-capita household expenditure is included as an indicator of the household's spending power (for developing countries, expenditure is regarded as a more reliable and accurate measure of well-being that can also proxy for income).

In line with the theories of Modigliani & Brumberg (1954) and Milton Friedman (1957), income smoothing could play an important role in consumption patterns. The age and the years of education of the household head influence their earning capacity; that is, they act as further proxies for permanent income (Charles et al., 2009:432). Intuitively, one would expect that a household with a higher expectation of future earnings may consume more than other households currently earning the same income.

Modelling the share of expenditure devoted to conspicuous consumption as the dependent variable should control for existing levels of asset ownership, which are expected to vary between groups. According to the asset deficit hypothesis, the observed racial differences in consumption expenditure patterns between groups are expected to decrease substantially over time because they are dictated by asset accumulation patterns – over lifecycles, but also across generations. Most members of the black middle class are young and in many cases they are the first generation of their families to belong to the middle class. Compared with both their older counterparts and those born into middle-class families (e.g. many white members of the middle class), these younger middle-class households start their careers with a deficit of assets. Within their financial constraints, such households can only eliminate their asset deficit gradually and over an extended period.

Table 2 presents the factors associated with a higher conspicuous consumption share for middle-class South Africans. Because of the correlation between household expenditure per capita and the educational attainment of the head of the household, four sets of results are presented, with educational attainment and the asset index being excluded in some regressions. Each regression shows a basic model of spending, including proxies

Table 2: Regression analysis of conspicuous consumption

Variable	(1)	(2)	(3)	(4)
Log expenditure per capita	−0.01523	−0.00891	−0.03201**	−0.02549
	(0.01546)	(0.01560)	(0.01600)	(0.01615)
Log expenditure per capita squared	−0.00005	−0.00033	0.00090	0.00061
	(0.00088)	(0.00088)	(0.00091)	(0.00091)
Age of household head	−0.00096***	−0.00093***	−0.00099***	−0.00095***
	(0.00009)	(0.00009)	(0.00009)	(0.00009)
Rural	−0.00652**	−0.00744**	−0.00551*	−0.00633*
	(0.00325)	(0.00326)	(0.00329)	(0.00330)
Number of children in household	−0.00615***	−0.00507***	−0.00613***	−0.00510***
	(0.00119)	(0.00125)	(0.00121)	(0.00126)
Number of elderly people in household	−0.00337	−0.00346	−0.00315	−0.00329
	(0.00214)	(0.00214)	(0.00215)	(0.00215)
Female-headed household	0.00315	0.00347	0.00309	0.00343
	(0.00215)	(0.00215)	(0.00215)	(0.00215)
Coloured	0.01335***	0.01293***	0.01346***	0.01312***
	(0.00350)	(0.00350)	(0.00351)	(0.00351)
Indian	−0.00932**	−0.00929**	−0.00936**	−0.00929**
	(0.00416)	(0.00416)	(0.00415)	(0.00415)
Emerging black middle class	0.03358***	0.02835***	0.03323***	0.02800***
	(0.00287)	(0.00341)	(0.00289)	(0.00342)
Established black middle class	0.01526***	0.01540***	0.01504***	0.01509***
	(0.00300)	(0.00299)	(0.00301)	(0.00301)
Asset index		−0.00437***		−0.00447***
		(0.00153)		(0.00156)
Education attainment of household head			−0.00048	−0.00030
			(0.00039)	(0.00040)
Constant	0.24592***	0.21110***	0.32595***	0.28857***
	(0.06851)	(0.06953)	(0.07104)	(0.07218)
Observations	4 392	4 392	4 324	4 324
R-squared	0.21419	0.21565	0.22115	0.22262

Notes: Standard errors in parentheses. ***$p < 0.01$, **$p < 0.05$, *$p < 0.1$.
Source: Authors' own calculations from IES.

for spending power (i.e. expenditure per capita and its square) and a range of demographic characteristics that can drive consumption priorities and preferences (i.e. race and household structure). A dummy for a rural location is included because rural inhabitants often face different prices and choice sets than urban inhabitants. Charles et al. (2009) and Kaus (2013) define reference groups as a composite of race group and region; since sufficiently comprehensive district level data do not exist in the IES, the regressions simply control for the urban/rural distinction. We include both the log of per-capita expenditure and the square of log per-capita expenditure because the relationship between logged household per-capita expenditure and conspicuous consumption as a share of expenditure is an inverted U, rising steeply at low levels of expenditure and then flattening out and eventually declining at higher levels of expenditure.

All of the coefficients have the expected signs. Expenditure per capita and expenditure per capita square both have negative coefficients, showing that for the South African middle class the share of expenditure allocated to conspicuous consumption decreases with an increase in expenditure per capita. As reported earlier, this negative relationship only holds at the top end of the expenditure distribution and there is a positive relationship between conspicuous consumption share and expenditure at lower levels of expenditure. This is consistent with a view in which the upper classes, proxied here by households with high expenditure, do not experience the same need to signal their wealth as the middle classes.

Age has a negative coefficient, showing that younger members of the middle class tend to allocate a greater share of expenditure towards conspicuous consumption. Given that separate proxies for expenditure, education and assets have been included, the negative coefficient on age may be interpreted as a possible indication that younger members of the middle class may still feel more vulnerable, perhaps because they have fewer years of experience in the labour market. This may lead to them being more easily elicited to signal their wealth through their consumption behaviour. They are also more likely to have recently arrived in the prosperous group, particularly if they are not white.

Rural inhabitants tend to have a lower conspicuous consumption share, which could perhaps be attributable to lower social pressure in rural areas; that is, the reference group may differ.

As would be expected, households with more dependents (children and elders) are less prone to conspicuous consumption. Put differently, where a greater share of the household consists of adults of working age, a higher conspicuous consumption share would be expected. This would also be associated with higher disposable income in households with more productive capacity.

The coefficient on female-headed households is positive, but not significant. A positive association was expected due to such households representing a highly vulnerable segment of the population.

The white middle class is the reference group for the regression. Accordingly, a positive coefficient on the racial indicator variable shows that members of the relevant racial group tend to devote a higher proportion of expenditure to conspicuous consumption than the white middle class. The work of Charles et al. (2009) and Kaus (2013) would predict that the urge to signal wealth would be stronger amongst coloured and black South Africans, who experience higher intra-group inequality and lower mean income. Consequently, prosperous members of these groups would be more inclined to use visual cues to set themselves apart from the rest of their reference groups. These results confirm this and show that, all other things equal, coloured and black households spend a significantly larger share of their money on items that can visibly signal wealth.

Note, however, that the coefficient on the black emerging middle class indicator variable is almost double that of the coefficient on the black established middle class. This suggests that a large part of observed black conspicuous consumption is driven by the emerging middle-class group, which appears to have a greater signalling need even after controlling for income levels. This may be associated with recent arrival in or due to uncertain continued membership of this status class, due, for instance, to relatively low educational attainment.

In line with the asset deficit hypothesis, the asset index is significant and negative and reduces the coefficient on the black emergent indicator variable. Interestingly, Figure 1

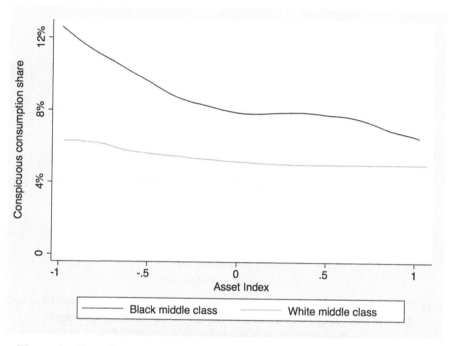

Figure 1: Conspicuous consumption shares of black and white middle-class households by asset index
Source: Authors' own calculations from IES.

shows that the black middle class' conspicuous consumption shares converge towards white middle-class levels as their asset levels increase. The wide divergence in conspicuous expenditure at low levels of assets may suggest that, for first-generation middle-class members, the asset accumulation process runs parallel to and proxies for equally important socio-economic orientation and consolidation processes. These may include the expansion and deepening of social networks, gaining more labour market experience and tacit knowledge about social conventions and systems. Such anchoring and rooting processes would enhance feelings of security and belonging and objectively reduce vulnerability through building social capital, which would then reduce the need to signal wealth via conspicuous consumption.

Whilst black middle-class households spend more on conspicuous consumption as a share of total expenditure, it is equally true that their spending patterns reflect that they are still catching up on household asset stocks. This is particularly true for the emerging group amongst black middle-class households. In other words, there does not appear to be a significant trade-off for these households between conspicuous consumption expenditure and filling the asset deficit. To augment regression analysis, Figure 2 shows that middle-class households with lower stores of assets (i.e. emerging middle class) were more likely to report recent asset purchases (over the previous 12 months) than their established counterparts. The only exceptions are cars and televisions. Regarding assets such as these where major innovations occur frequently, consumers may replace goods earlier than strictly required in order to signal social status or to access productivity gains or luxury features associated with the additional functionality of the new version. For most assets, however, the emerging black middle

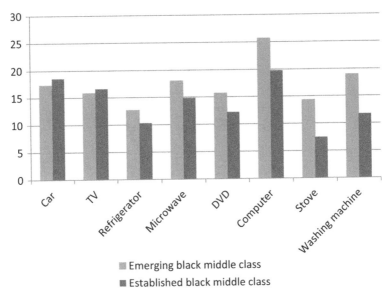

Emerging black middle class
Established black middle class

Figure 2: Probability of having purchased certain assets in the previous 12 months for the emerging and established black middle class (percentages)
Source: Authors' own calculations from IES.

class is considerably more likely to have reported recent purchases. The strikingly high incidence of computer purchases amongst the emerging black middle-class group is particularly interesting because it may speak more directly to the socialisation associated with recently joining the middle class, referenced earlier.

6. Conclusions

The results of the empirical research reported here are consistent with the hypotheses set out earlier. Conspicuous consumption does appear to be socially contingent, where the reference group is defined as a particular race group. Such conspicuous consumption increases in groups where mean incomes are lower and inequality higher, as was predicted and also found in earlier studies in both South Africa and the USA. Also, in line with hypotheses, a large proportion of the observed high conspicuous consumption levels within the black middle-class group can be explained by new or perhaps insecure membership of the group. Finally, the empirical evidence shows that conspicuous consumption is negatively related to asset ownership. Thus, as asset ownership rises, it is likely that conspicuous consumption decreases – perhaps because the need to signal economic status declines commensurately.

The available evidence thus supports an explanation for the pattern of spending amongst black middle-class consumers being different from that of their white counterparts due to two factors: their vulnerability as new entrants to the middle class, and their associated asset deficit. Conspicuous consumption is natural amongst those who are economically more successful than the mean within their reference group, particularly when this group is not prosperous on average. It is likely to be a more prominent feature of consumption whilst this higher economic status is still new, and also where such status is still tenuous because of real or perceived income vulnerability.

If the hypotheses set out here hold, one would expect the more established part of the black middle class to grow over time and to exhibit consumer patterns that would increasingly resemble those of other established members of the middle class. If economic trends continue fuelling growth of the black middle class, social mobility will propel more individuals into the middle class, refilling the ranks of the emergent middle class. Conspicuous consumption will thus be an enduring feature of South African consumer expenditure patterns.

Given that both this work and that of Kaus (2013) are reliant on the IES data, it will be useful to test the robustness of these findings with analysis from other data sources such as the All Media and Products Surveys.

References

Alexander, P, 2010. Rebellion of the poor: South Africa's service delivery protests – A preliminary analysis. Review of African Political Economy 37(123), 25–40.

Bourdieu, P, 1984. Distinction. A Social Critique of the Judgement of Taste. Routledge, London.

Booysen, F, van der Berg, S, Burger, R, von Maltitz, M & Du Rand, G, 2008. Using an asset index to assess trends in poverty in seven sub-saharan african countries. World Development 36(6), 1113–1130.

Charles, KK, Hurst, E & Roussanov, N, 2009. Conspicuous consumption and race. Quarterly Journal of Economics 124(2), 425–67.

Easterly, W, 2001. The middle class consensus and economic development. Journal of Economic Growth 6, 317–35.

Friedman, M, 1957. A Theory of the Consumption Function. Princeton University Press, Princeton, NJ.

Goldthorpe, JH & McKnight, A, 2004. The economic basis of social class. CASE paper, 80. Centre for Analysis of Social Exclusion, London School of Economics and Political Science, London.

Kaus, W. 2013. Conspicuous consumption and 'race': Evidence from South Africa. Journal of Development Economics 100(1), 63–73.

Krige, D, 2011. Debating the black diamond label for South Africa's black middle class. Paper presented at the annual Anthropology Southern African Conference, 3–5 September, Stellenbosch University.

Lamont, M & Lareau, A, 1988. Cultural capital: Allusions, gaps, and glissandos in recent theoretical developments. Sociological Theory 6, 153–68.

Lopez-Calva, LF & Ortiz-Juarez, E, 2011. A vulnerability approach to the definition of the middle class. Policy Research Working Paper Series 5902. World Bank, Washington, DC.

Mamabolo, K, 2013. 4 Million and Rising – The South African Black Middle Class. www.fbreporter.com Accessed 5 November 2013.

Mbeki, T, 1998. Statement of Deputy President Thabo Mbeki at the Opening of the Debate in the National Assembly on 'Reconciliation and Nation Building', 29 May, National Assembly, Cape Town.

Modigliani, F & Brumberg, R, 1954. Utility analysis and the consumption function: An interpretation of cross-section data. In Kurihara, KK (Ed.), Post-Keynesian Economics. Rutgers University Press, New Brunswick, NJ.

Nieftagodien, S & Van der Berg, S, 2007. Consumption patterns and the black middle class: The role of assets. Working Paper 02/2007, Stellenbosch University, Department of Economics

Office of the President, 2013. Development Indicators 2012. The Presidency, Pretoria. http://www.thepresidency.gov.za/MediaLib/Downloads/Home/Publications/DPMEIndicators2013/DPME%20Indicators%202013.pdf Accessed 8 November 2014.

Ravallion, M, 2010. The developing world's bulging (but vulnerable) middle class. World Development 38(4), 445–454.

Schlemmer, L, 2005. Lost in transformation? South Africa's emerging African middle class. CDE Focus No 8, August, Centre for Development and Enterprise, Johannesburg.

Trigg, AB, 2001. Veblen, Bourdieu and conspicuous consumption. Journal of Economic Issues 3(1), 99–115.

Veblen, T, 1899. The theory of the leisure class. In the collected works of Thorstein Veblen Vol. 1. Reprint. Routledge: London.

Visagie, J & Posel, D, 2013. A reconsideration of what and who is middle class in South Africa. Development Southern Africa 30(2), 149–67.

Life chances and class: Estimating inequality of opportunity for children and adolescents in South Africa

Asmus Zoch[1]

This paper aims to determine the degree to which class and socio-economic background influence a child's life chances and their future perspectives. We build on the growing number of papers that deal with the concept of inequality of opportunity. Comparing children from lower and upper middle-class households we find significant differences in terms of access to basic education, sanitation, clean water and mobility. Our multivariate analysis highlights the importance of class membership for schooling outcomes and labour market prospects of a child. The single most important variable to explain schooling outcomes is mother's education. While income seems to be less important for younger ages, it becomes increasingly important for the chances of reaching matric and obtaining tertiary education. The results are robust for various models and panel data.

1. Introduction

According to Du Toit & Kotzé (2011:77), 'formal equality in South Africa was achieved by constitutional ruling, but actual socioeconomic quality, however, was not'. This simple statement points to one of the core problems in South African society after the political transition, namely the lack of substantive economic inclusion following political liberalisation.[2] Almost two decades since the end of apartheid, poverty has retained its strong racial dimensions. This remains the state of affairs despite the elimination of formal discriminatory rules and legislation.

While poverty has remained very concentrated amongst blacks, there has however been some evidence of improved social mobility at the top end of the income distribution, attested to by an increase in the black share of affluence. Consequently, there has been rising inequality within the racial groups, with a sharp increase in the Gini coefficient of both the white and the black population groups in the post-apartheid period (Leibbrandt et al., 2010; Yu, 2010; Van der Berg, 2011).

There is comprehensive literature about inequality and poverty in South Africa, including income inequality, poverty (Hoogeveen & Özler, 2006; Leibbrandt et al., 2010; Özler, 2007) and education inequality (Lam et al., 2011, Van der Berg, 2007, 2009). Other literature dealt with the concept of classes and the rising black middle class (Seekings & Nattrass, 2005; Gornick & Jäntti, 2013). While the concept of

---

[1]PhD Student, Department of Economics, University of Stellenbosch, Private Bag X1, 7602 Matieland, South Africa.
[2]This statement refers to the public discourse claiming that participation in the labour market and a movement out of unemployment has not been realised.

inequality or class is rather static, this paper uses the more dynamic concept of inequality of opportunity. This approach has the advantage that it gives a prediction of future inequality by looking at inequality of opportunity within children. Even within a very unequal society, inequality is more tolerable if opportunities are equal to future generations.

This paper adds to the existing literature by combining the concept inequality of opportunity and class. While there are some studies looking at inequality of opportunities in terms of service delivery (e.g. World Bank, 2012), this paper more explicitly looks at how parental background and class influence opportunity of their offspring. Such a focus is insightful and useful because studies have shown that gaps which emerge early in life tend to be permanent and often widen further over the lifecycle (Heckman, 2000). Additionally, this focus may also be strategic because there is an increasing body of evidence in support of early invention and this has also garnered policy prominence in South Africa and internationally. This paper evaluates access to opportunities in South Africa by comparing the opportunities of individuals from different socio-economic backgrounds at different life stages, in terms of access to basic amenities and goods such as clean water and food at a young age, reaching matric or obtaining tertiary education, and finding employment as an adult. The paper shows that class and socio-economic background have a dramatic influence on a child's life chances in terms of these criteria and that such opportunities affect a child's starting point in life. Our multivariate analysis highlights the importance of class membership for schooling outcomes and labour market prospects of a child.

2. Theory and literature review

This section of the paper provides a short review of the literature and the concept of inequality of opportunity. In the past two decades, several studies have observed socio-economic and intergenerational mobility; for example, Corak (2006) has shown in an international comparison that children from low-income families become low-income earners. More recently, other researchers have focused attention on the conceptual issue of mobility and on the notion of equality of opportunity (e.g. Paes de Barros et al., 2009; Bourguignon et al., 2007; Ferreira & Gignoux, 2011). For South Africa, a paper by Burns & Keswell (2011) focuses on intergenerational persistence of educational status in KwaZulu-Natal, and finds that such persistence has increased over generations, while a paper by the World Bank (2012) observes inequality of opportunity among children in South Africa using the Human Opportunity Index (HOI). Another paper comparing two different measures for inequality of opportunity in a number of different countries including South Africa by Brunori et al. (2013) concludes that most differences in inequality of economic opportunity arise from exogenous factors such as family background, race or gender and not due to individual effort. Finally, a further paper by Piraino (2012) developed a method to measure inequality of opportunity in South Africa.

Most of the research regarding inequality of opportunity is motivated as a theory of fairness and justice. Commencing with Rawls's (1971) *A Theory of Justice* and Sen's (1980) *Tanner Lectures*, political philosophers and economists have discussed the appropriate sphere wherein equality should be promoted. A key development in this discussion is the incorporation of a central role for personal responsibility into the definition of fairness (Ferreira et al., 2011); that is, equality of opportunity is often described as an ideal compromise between different perspectives on equity, because it retains the vital dimensions of egalitarianism, while simultaneously allowing for differences in outcomes based on effort, merit and other relevant criteria. Allowing outcomes to differ is critical because this entails that choices can have consequences, and consequences can in turn help to inform and reward individuals' behaviour.

This analysis will follow the approach by John Roemer (1998), who acknowledges this tension between egalitarianism and effort by differentiating between two potential sources of unequal outcomes, namely circumstances (factors exogenous to the person, such as gender, race, family background or place of birth) and individual efforts (outcome determinants that can be affected by individual choice). Essentially, this forms the basis of a simple binary view of the fairness of life chances. Within this conceptual framework, a level playing field is one where one's fate is largely determined by one's own efforts, rather than being determined by inherited factors and circumstances such as one's family background, gender or race.

Our approach acknowledges the New-Weberian school, in as much as we wish to ascertain the extent to which factors such as inequality of life chances among individuals and families are structured on the basis of class. One central claim in the New-Weberian tradition is that variations in market positions arise out of differences in the possession of market-relevant assets that determine life chances. Inspired by this work, instead of merely investigating how life chances limit choices, we invert the question and investigate how class influences the ability to escape from poverty and enables the next generation to form their own choices.

3. Data

The two panel studies used in this paper are KwaZulu-Natal Income Dynamics Study (KIDS) and National Income Dynamics Survey (NIDS). Most of the analysis conducted in this paper comes from the first wave of NIDS. The first representative national panel study, conducted by the Southern Africa Labour and Development Research Unit at the University of Cape Town, NIDS is a large and representative survey, with 31 163 individual observations and 6921 households in 2008. The survey includes detailed information about living conditions, education, household formation, occupation, income and expenditure. In addition, it includes various questions on subjective well-being and satisfaction levels. Consequently, we can use this dataset to compare the living conditions, education and labour market outcomes as well as self-perception for a representative sample of youths, coming from different class backgrounds.

The only disadvantage of NIDS is that it covers a relatively short time period, spanning four years. For this reason, we also test our findings using KIDS, a three-wave panel dataset spanning the first decade of South Africa's democracy. However, KIDS only covers the province of KwaZulu-Natal and is limited to black and Indian households, thereby excluding households with coloured or white heads due to likely sampling bias (see Agüero et al., 2007). However, Woolard & Klasen (2005) show that the African population in KwaZulu-Natal are comparable with Africans elsewhere. Overall attrition is reasonable, with 1132 households (83.6%) having been successfully re-interviewed for the second wave in 1998 (Adato et al., 2006). For the third wave in 2004, some 74% of the households contacted in 1998 were re-interviewed. In total, 841 households could be successfully interviewed through all three survey waves. Having a time span of 11 years allows one to test for the robustness of the results from NIDS dataset and to control for potential endogeneity.

4. Estimating inequality of opportunity at various life stages

In this section of the paper we will examine life chances by comparing the respective opportunities of young school children, emerging adults and young adults from different socio-economic classes and backgrounds. Comparing the opportunities of young children from privileged and disadvantaged backgrounds is informative because young individuals have had limited opportunity to differentiate themselves based on effort and hard work, thus differences in opportunities between them are largely attributable to differences in the circumstances into which they were born. Further, the level of education is of particular interest due to its strong correlation with life chances (Burns & Keswell, 2011). We believe that looking at emerging adults and their school to work transition is particularly of interest in the case of South Africa where youth unemployment is exceptionally high.

4.1 Circumstances and opportunities of young children

Access to key goods and services such as clean water, basic education, health services, minimum nutrition and citizenship rights is crucial to allow individuals to pursue a life of their own choosing (World Bank, 2010:32). Providing children with a complete set of basic opportunities is essential in affording them the opportunity to make their own life choices and to realise their productive potential.

Figure 1 considers the likelihood that children aged between 10 and 14 would live in certain circumstances and would reach various benchmarks based on their socio-economic background. To this end, we compare the life chances of a child from lower-class households with those of upper middle-class households. For the purposes of the subsequent analysis, a household is defined as 'lower class' if it belongs to the bottom two quintiles, and neither parent has completed primary education. A household is classified as 'upper middle class' if it belongs to the richest income quintile, at least one biological parent live in the household and at least one parent has completed high school or achieved a higher education. Choosing the cut-off lines for the class analysis is somehow arbitrary since there are many different ways to define the 'middle class' in the literature (Hertova et al., 2010) and different approaches do not identify the same people as middle class in South Africa (see Visagie & Posel, 2013). We follow the approach of Easterly (2001) and Barro (1999) and other studies and use income deciles as well as parent education.

In most cases – apart from reaching Grade 7 on time[3] – these circumstances and benchmarks are not under the direct control of these children and differences should therefore be interpreted as indicative of inequalities that cannot be attributed to effort.

The results[4] show that living in a disadvantaged household reduces the chances of reaching Grade 7 on time by about 36 percentage points (52% vs. 88%). Children completing Grade 7 on time are more likely to have had access to schools of reasonable quality and can avoid unnecessary grade repetition (World Bank, 2010:45). However, some provinces in South Africa do not endorse grade repetition, which might lead to an underestimation or mis-estimation of inequality of school quality when using this measure.

---

[3] Which is defined as being in the right class for a child's age.
[4] The estimations reported are based on NIDS 2008.

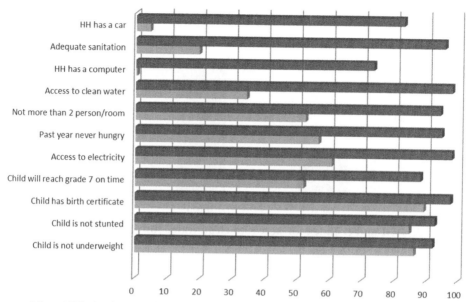

Figure 1: Estimated probability of reaching various benchmarks for children from different socio-economic background (age 10 to 14)
Note: HH, household.

The outcome differences are even more pronounced in terms of the chances of a child from a poor socio-economic background living in a household with access to adequate sanitation (20% vs. 96%) and clean water (34% vs. 98%).[5] Having access to clean water and sanitation are important factors in determining susceptibility to serious illnesses like cholera. For example, Mugero & Hoque (2001) find that contaminated water and low latrine coverage are key risk factors for a cholera epidemic in affected rural areas. The chances of having access to a car (4% vs. 83%) or a computer (0.4% vs. 73%) are close to zero for children from lower class background. On the other hand, there seem to be only small differences in the likelihood of the child being underweight (six percentage point difference) or being stunted (eight percentage point difference) between children from lower and upper middle-class households.

In terms of living conditions, the likelihood that a child will live in an overcrowded household (more than two people per room) and with no electricity (a key factor enabling a child to read and do homework when it is dark) is between 38 and 42 percentage points higher when the child is from a lower class background. These children also face a much higher likelihood of being hungry. Our results are in line with the findings of the Word Bank (2012) study: for the chosen set of basic goods we find significant differences in the access for children from different backgrounds.

As Figure 1 shows, socio-economic class has a very marked influence on the opportunities and circumstances of children. The literature suggests that these

---

[5] Access to clean water is defined as having a piped water tap in the dwelling, on site or in the yard. Clean sanitation is defined has having either a flush or chemical toilet.

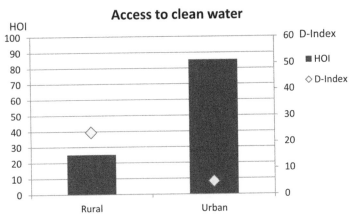

Figure 2: HOI and D-Index for access to clean water for children (age 10 to 14)
Source: Author's own estimation based on NIDS 2008.

circumstances and opportunities mean that young children from disadvantaged and deprived households have a very different starting point in life than young children from upper middle-class households.

To further examine the origins of the observed differences, we adopt the HOI used in several studies by the World Bank (2010, 2012) to explore inequality of opportunity for children. The HOI measures in a single indicator the absolute coverage rate of a particular service, adjusted by how equitably the available services are distributed among groups differentiated by circumstance (World Bank, 2012:18). The first part, the average coverage rate, can be directly adopted using household survey data. The second part uses a dissimilarity index (D-Index), which measures the dissimilarity of access rates for a given service for groups defined by the circumstance characteristics (race, location, gender, etc.), compared with the average access rate of the whole population (World Bank, 2010).

Figure 2 shows the HOI and the D-Index for access to clean water. While a HOI of 100 points would indicate universal access, zero points mean that there is no access to clean water at all. The HOI bar for urban children has a value of about 85 points and the D-Index is about five points. Therefore, water coverage rates are high in urban areas and the differences in opportunities are not based on circumstances. In contrast, the HOI has only a value of about 25 points, indicating much lower access rates in rural areas. In addition, the D-Index has a value of 40 points which indicates that about 40% of opportunities are inequitably allocated among circumstance groups.

Figure A1 in Appendix A shows significant differences for all infrastructure services in access to the services and the inequality within circumstance groups. Especially stark are the results for access to proper sanitation, which is nearly absent in rural areas and very unequally distributed. While the possession of cars is highly influenced by their circumstances, there are no major differences between children from rural and urban areas. The same patterns hold for computers. These differences in possessions can be partly explained by an asset deficit of the black population who could not accumulate wealth during the apartheid period. Finally, for rural children, finishing Grade 7 on time is less likely and more dependent on background.

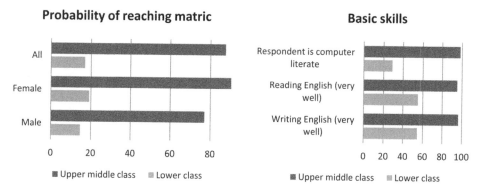

Figure 3: Education differences for emerging adults by socio-economic background (age 19 to 20)

The next section focuses on the next stage in people's lives, examining how socio-economic class and background affect outcome variables such as reaching matric, accessing tertiary studies and securing employment in later adolescent life stages.

4.2 Schooling, skill achievement and life satisfaction in emerging adulthood

Figure 3 illustrates that the likelihood of an emerging adult[6] reaching matric by the age of 19 or 20 vastly differs between those from a poor socio-economic background and those from upper middle-class households (17% vs. 88%). For male students from disadvantaged backgrounds, the probability is lower than 15%. In addition, there are significant differences in the self-reported ability to read and write English competently and in computer literacy. Such skills are very likely to be relevant for the individual's labour market prospects.

Figure 4 suggests that the disparities in the opportunities and circumstances of these emerging adults affect their life satisfaction. On a satisfaction scale (from one to 10, 10 being most satisfied), emerging adults with upper middle-class parents report considerably higher average satisfaction levels than those with lower class parents. These differences could be attributed to anticipated life circumstances, their current living conditions or a combination of the two. It is telling that such stark differences in reported happiness and life satisfaction can already be discerned at such early stages of these individuals' lives.

4.3 Young adults and labour market prospects

Figure 5 illustrates the chance a young adult (aged 21 to 25) has of reaching matric, achieving a tertiary qualification and obtaining employment (left-hand panel). Again, those from a lower class household have a considerably lower likelihood of reaching matric by this age than those with upper middle-class parents (19% vs. 77%). Similarly, the likelihood of obtaining some tertiary education is extremely low (1% vs. 50%). Consequently, these individuals also face a much higher likelihood of not finding employment (46% vs. 15%). Yet young adults from an upper middle-class

---

[6]The term 'emerging adult' was coined by Arnett (2000) for late teens to mid-twenties. In this analysis we used the age group 19 to 20.

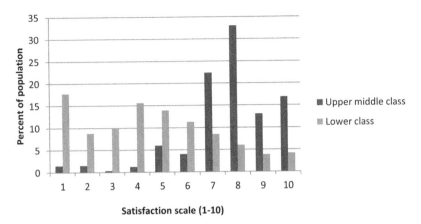

Figure 4: Life satisfaction of emerging adults from different parental class (age 17 to 20)
Source: Author's own calculation based on NIDS 2008.

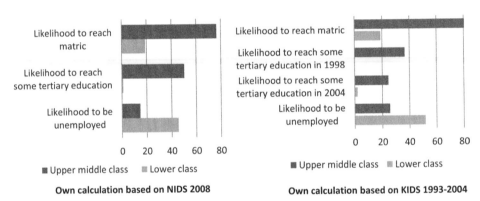

Figure 5: Likelihood of a young adult (21 to 25) reaching matric, obtaining some tertiary education and not finding employment

background are more likely to be attending university, which might overestimate the difference. However, they will have a higher likelihood of being offered employment with the completion of their studies.

The first wave of NIDS in 2008, however, can only provide a description and no causal correlations. For this reason, we test the robustness of these findings using KIDS 1993–2004. Here we estimate the likelihood that, dependent on the socio-economic class of their parents, a child aged between seven and 17 in 1993 will reach matric or some tertiary education in 2004 or 1998. The results correspond with NIDS findings: a child who lived with lower class parents in 1993 has only a 19% chance of reaching matric in 2004, while a child with upper middle-class parents has an 82% chance. Additionally, the analysis shows that a child with lower class parents has a near-zero chance of accessing tertiary education.

While these results appear to confirm and support the results from the cross-section, we need to note that younger household members who left the household were not followed by KIDS panel, and therefore children leaving the household to pursue tertiary

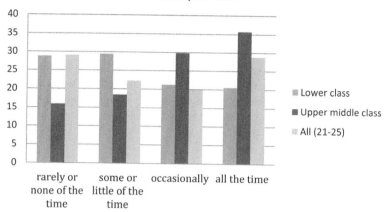

Figure 6: Respondent's future perspective (age 21 to 25)
Source: Author's own calculation based on NIDS 2008.

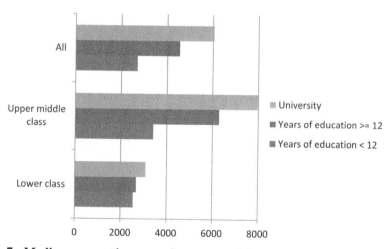

Figure 7: Median reservation wage for young adults (age 21 to 35)
Source: Author's own calculation based on NIDS 2010.

opportunities may often disappear from the sample. This may also help explain why KIDS shows that fewer students accessed tertiary education in 2004 than in 1998.

Finally, Figure 6 illustrates the level of optimism reported by respondents of this age group regarding their future. Here again, we detect notable differences between the reported optimism of young adults from lower and those from upper middle-class households.

Another indicator of self-perception that further highlights the role of socio-economic background is the reservation wage: the lowest wage at which people would be willing to work in the labour market. In NIDS 2010 the reservation wage is identified by the following question: 'What is the absolute lowest monthly take-home wage that you would accept?'. There are some concerns (e.g. Kingdon & Knight, 2001) that responses to reservation wage questions in survey data are not very reliable. However, since the regression results are in line with economic theory, we believe responses to reservation wages and true economic behaviour are significantly correlated.

As illustrated in Figure 7, coming from a poor socio-economic background reduces the reservation wage of young adults quite dramatically. While young adults with an upper middle-class background who went to university have a median reservation wage of about R12 000, the median reservation wage is only about R3000 for their lower class counterparts. This might be explained either by young adults with a better socio-economic background being able to afford to wait for a high-paying job, or that they have a better self-perception of their true 'market value', since they are able to compare themselves with their parents.

4.4 Regression analysis

Identifying the factors that explain differences in early school performance is important to understand life opportunities and social mobility. Without access to a good education, children's choices will be severely constrained. There are two plausible ways through which parental education can influence their children's school performance: a direct effect through inherited genetics (intergenerational ability), and an indirect effect via support, assistance and encouragement (parents with higher education will value school quality and education more highly and are able to help their children to perform well at school) (Buis, 2012).

In Table 1, column (1) illustrates the results of a linear regression estimating the likelihood of reaching Grade 7 on time.[7] Note the convex relationship indicated by the significance of the coefficient on the squared value of the mother's education; the father's education has a positive coefficient but is not significant. Hence, a mother's education seems to have a significantly greater influence than a father's education, a finding that corresponds to other studies (Lam et al., 2011; Burns & Keswell, 2011; Buis, 2012). If parental education works either through helping children perform well at school or through the transmission of cultural capital, one would expect that a mother's education has a stronger effect, since the mother traditionally spends most time with the children (Buis, 2012:6).

The model also shows that on average female students perform significantly better than male students. The same is true for children from a coloured and Indian background. On average, white students do not have a better school record in early life stages, ceteris paribus. This means black children have the same grade progression in early life stages if they have similar socio-economic background, something that often does not apply.[8] The number of children in a household is negatively correlated with the likelihood of reaching Grade 7 on time, whereas household income has a positive but insignificant impact (richer parents may send their children to better quality schools, which may not be reflected in school repetition by Grade 7).

Column (2) of Table 1 illustrates that the education levels of parents are positive and significantly correlated with the likelihood of reaching matric by the ages of 19 to 20.

---

[7] A linear model is used since it can be straightforwardly interpreted. As a robust check we also run a logit model (see Table A1 in Appendix A).
[8] Note that the outcome variable only measures school attainment despite differences in the quality of education the average black and white child receives. This does not mean increasing the average income of black households will solve all school problems in South Africa. We know that there are strong school fixed effects and that the traditional black part of the school system is significantly underperforming.

Table 1: School outcome regressions

Variable	(1) Age 10 to 14, Grade 7	(2) Age 19 to 20, matric
Mother lives in the household	0.0233	0.0184
	(0.0261)	(0.0421)
Mother's years of education	−0.00386	−0.0291**
	(0.00811)	(0.0117)
Mother's years of education^2	0.00171***	0.00337***
	(0.000586)	(0.000938)
Father lives in the household	0.0604**	0.0561
	(0.0273)	(0.0471)
Father's years of education	0.000227	0.0109**
	(0.00275)	(0.00457)
Log of per-capita household income	0.0202	0.0385*
	(0.0128)	(0.0226)
Female household head	0.0115	0.0918**
	(0.0233)	(0.0425)
Number of children	−0.0159**	−0.0148
	(0.00634)	(0.00925)
Coloured	0.0877**	−0.150**
	(0.0414)	(0.0754)
Indian	0.148***	−0.314***
	(0.0505)	(0.110)
White	−0.0298	0.0217
	(0.0611)	(0.0930)
Female	0.180***	0.108***
	(0.0201)	(0.0365)
Tribal	0.0207	−0.0963
	(0.0425)	(0.0694)
Urban formal	0.0535	0.0860
	(0.0442)	(0.0758)
Urban informal	0.0510	0.0233
	(0.0589)	(0.0920)
Constant	0.320***	−0.0938
	(0.107)	(0.177)
Observations	3 305	1 157
R-squared	0.138	0.273

Note: Based on NIDS 2008. Not listed: dummy variables for provinces and if mother and father education observations were missing. Robust standard errors in parentheses. ***$p < 0.01$, **$p < 0.05$, *$p < 0.1$.

Again, the mother's education squared term is highly significant and positive, indicating convex returns on a mother's education.

In comparison with the first regression, it is notable that per-capita household income is now larger and significant. Therefore, while in the early life stages education seems to be heavily dependent on the mother's education, income becomes more important in explaining differences in later outcomes of the school career.

Table 2: Tertiary education regression

Variable	(1) Some tertiary education	(2) Further education (except university)	(3) University
Mother's years of education	−0.0158*	−0.00747	−0.0196***
	(0.00823)	(0.00774)	(0.00638)
Mother's years of education^2	0.00199***	0.00132*	0.00190***
	(0.000672)	(0.000674)	(0.000533)
Father's years of education	0.0111***	0.00896***	0.00358
	(0.00342)	(0.00312)	(0.00227)
Log of per-capita household income	−0.239***	−0.137*	−0.157***
	(0.0725)	(0.0833)	(0.0562)
Log of per-capita household income^2	0.0245***	0.0151**	0.0147***
	(0.00620)	(0.00726)	(0.00493)
Coloured	−0.0486	−0.0386	−0.0251
	(0.0407)	(0.0384)	(0.0280)
Indian	0.177	0.0541	−0.0479
	(0.144)	(0.174)	(0.0679)
White	−0.121*	−0.149**	0.0651
	(0.0700)	(0.0696)	(0.0618)
Female	0.0239	0.0125	0.0103
	(0.0201)	(0.0199)	(0.0127)
Tribal	0.0337	0.00713	−0.0163
	(0.0402)	(0.0480)	(0.0187)
Urban formal	0.103**	0.0302	0.0469*
	(0.0478)	(0.0560)	(0.0273)
Urban informal	0.124**	0.0635	0.0408
	(0.0536)	(0.0594)	(0.0267)
Constant	0.500**	0.246	0.429***
	(0.217)	(0.237)	(0.161)
Observations	2315	2248	2121
R-squared	0.233	0.142	0.217

Note: Based on NIDS 2008. Not listed: province dummies and whether or not parents are living in household. Robust standard errors in parentheses. ***$p < 0.01$, **$p < 0.05$, *$p < 0.1$.

The next analysis focuses on factors that determine whether or not a young adult will reach some tertiary education by the ages of 20 to 26. Since South Africa generally has strong convex returns on education, achieving tertiary education is an important factor associated with good prospects in the labour market. The results of Table 2, column (1) are in line with our earlier findings, namely the importance of parents' education and household income. Note the highly significant squared term on household income, indicating that richer households are much more likely to enable their children to get some tertiary education. These results stand in opposition to the findings of a study by Lam et al. (2010), who do not find strong credit restraints for South African students. However, their study is confined to students from the Western Cape participating in the Cape Area Panel Study, and regressions control for test results.

To test the robustness of our results, we use the panel option of NIDS to follow those students enrolled in school during 2007/08 and note who is making the transition into tertiary education. The results of the panel regression shown in Table A2 in Appendix A confirm that parents' education is one important factor. We also note that the coefficient for household income is twice the size in Table A2, column (2). Household income therefore seems to have a larger impact on the likelihood of

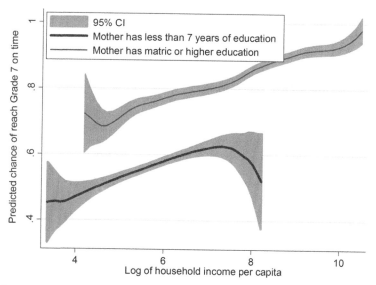

Figure 8: Correlation between mother's education/income quintile and reaching Grade 7 on time, derived from regression results (see Table 1)
Note: CI, confidence interval.

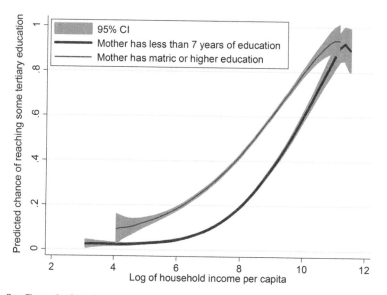

Figure 9: Correlation between mother's education/income quintile and reaching some tertiary education, derived from regression results (see Table 2)
Note: CI, confidence interval.

pursuing tertiary education than achieving matric. In addition, in Table 2 the coefficient for white is negative, which seems to be surprising. Yet, when splitting tertiary education into university and other forms of further education, the white coefficient is no longer negative or significant when looking at university education only.

Figures 8 and 9 show a graphic illustration of the model. In the first graph one can see a large increase in the probability of reaching Grade 7 on time (cf. reaching Grade 7 by the age of 14), when the mother has more than 12 years as opposed to less than seven years of education. This highlights the importance of mother's education, independent of the income of the household. Furthermore, the size of the confidence interval emphasises that there is much less unexplained variation when a mother has more than 12 years of education.

Table 3: Reservation wage regression

Variable	(1) All	(2) Employed	(3) Unemployed
Best age in years	0.0102**	0.0184***	0.00393
	(0.00399)	(0.00666)	(0.00547)
Years of education completed	−0.0311**	−0.00614	−0.0703***
	(0.0158)	(0.0167)	(0.0246)
Years of education completed^2	0.00644***	0.00462***	0.00879***
	(0.00102)	(0.00109)	(0.00162)
Mother's years of education	0.0127**	0.0210***	0.00572
	(0.00499)	(0.00797)	(0.00606)
Father's years of education	0.00306	−0.00706	0.0104*
	(0.00504)	(0.00822)	(0.00603)
Log of per-capita household income	0.121***	0.201***	0.0720**
	(0.0187)	(0.0312)	(0.0303)
Coloured	0.269***	0.250***	0.292***
	(0.0606)	(0.0727)	(0.1000)
Indian	0.150	0.235*	0.0753
	(0.0950)	(0.121)	(0.138)
White	0.360***	0.348***	0.272**
	(0.0861)	(0.105)	(0.133)
Urban formal	0.0876*	0.0626	0.148**
	(0.0528)	(0.0791)	(0.0650)
Urban informal	−0.0232	−0.0273	−0.0101
	(0.0610)	(0.0923)	(0.0793)
Tribal	0.0782	0.0996	0.0656
	(0.0498)	(0.0782)	(0.0634)
Female	−0.187***	−0.182***	−0.186***
	(0.0337)	(0.0523)	(0.0411)
Constant	6.172***	5.302***	6.784***
	(0.182)	(0.275)	(0.313)
Observations	5 323	1 984	3 345
R-squared	0.240	0.276	0.184

Note: Based on NIDS 2010. Outcome variable log (reservation wage). Not listed: dummies for mother and father education missing. Robust standard errors in parentheses. ***$p < 0.01$, **$p < 0.05$, *$p < 0.1$.

Figure 9 illustrates the influence that a mother's education and living in an upper middle-class household has on accessing tertiary education. The two graphs show that income explains much more of the likelihood of reaching tertiary than lower levels of education. For poor households (below R500 per-capita income), the chances of accessing tertiary education remain slim (below 10%) even with a mother that achieved 12 years of education. However, for individuals with per-capita income above R25 000, the probability of accessing a tertiary institution exceeds 90%, ceteris paribus. The figure also shows virtually no chance of reaching tertiary education for a child with a mother who has less than seven years of education and where the household's per-capita income is below R1000 per month.

Finally, Table 3 illustrates the results of a reservation wage regression to determine which factors influence the minimum income wage a young adult would work for. As expected, for South Africa there are convex returns to education, and therefore obtaining higher education will significantly increase the reservation wage. Secondly, the reservation wage is significantly higher for adults with well-educated parents. It is interesting to note that even for the reservation wage the mother's education seems to be more important than the father's. Most importantly, per-capita household income is significantly positive. While it seems obvious that while an individual is working his/her salary will determine the reservation wage, in column (3) we see that household income influences the decision even when the young adult is unemployed. In addition, it seems that coloureds and whites have significantly higher reservation wages, while females would work for a lesser amount than men, on average ceteris paribus.

5. Conclusion

This paper's aim was to determine the degree to which class and socio-economic background influence a child's life chances and their future perspectives. Using NIDS 2008 we found significant differences for children from lower and upper middle-class households in terms of access to sanitation, clean water, mobility, and so forth. In terms of their education, children are much more likely to reach Grade 7 on time when parents and especially mothers are well educated. While income seems to be less important for younger ages, it becomes increasingly important for the chances of reaching matric and obtaining tertiary education. Socio-economic background plays a major role in explaining who enters university and who does not. The results are robust for various models and panel data. Finally, our results show that people have much higher expectations of the labour market and therefore higher reservation wages when coming from a higher socio-economic background, even when controlling for education.

In conclusion, the findings of severe inequality of opportunity for South African youth show the urgency of further interventions. Studies have shown that early intervention at the childhood level might be most effective in increasing healthy physical, emotional, social and cognitive development (e.g. Heckman, 2006). Reducing inequality of opportunity will not only make sure South Africa becomes a more fair society, but will also reduce the inequality gap of future generations.

Acknowledgements

Many thanks go to Ronelle Burger and Servaas van der Berg for their helpful inputs and comments throughout this project.

References

Adato, M, Carter, M & May, J, 2006. Exploring poverty traps and social exclusion in South Africa using qualitative and quantitative data. Journal of Development Studies 42(2), 226–47.

Agüero, J, Carter MR & May, J, 2007. Poverty and inequality in the first decade of South Africa's democracy: What can be learnt from Panel Data from KwaZulu-Natal? Journal of African Economies 16(5), 782–812.

Arnett, JJ, 2000, Emerging adulthood: A theory of development from the late teens through the twenties. American Psychologist 55(5), 469–80.

Barro, R, 1999. Determinants of democracy. Journal of Political Economy, 107(6), 158–183.

Bourguignon, F, Ferreira, F & Menéndez, M, 2007. Inequality of opportunity in Brazil. Review of Income and Wealth 53(4), 585–618.

Brunori, P, Ferreira, FH & Peragine, V, 2013. Inequality of opportunity, income inequality and economic mobility: Some international comparisons. Policy Research Working Paper 6304, The World Bank, Development Research Group.

Buis, ML, 2012. The composition of family background: The influence of the economic and cultural resources of both parents on the offspring's educational attainment in the Netherlands between 1939 and 1991. European Sociological Review 29(3), 593–602.

Burns, J & Keswell, M, 2011. Inheriting the future: Intergenerational persistence of educational status in KwaZulu-Natal. Economic History of Developing Regions 27(1), 150–75.

Corak, M, 2006. Do poor children become poor adults? Lessons from a cross-country comparison of generational earnings mobility. Dynamics of Inequality and Poverty. Research on Economic Inequality 13, 143–88.

Du Toit, P & Kotzé, H, 2011. Liberal democracy and peace in South Africa: The pursuit of freedom as dignity. Palgrave Macmillan, New York.

Easterly, W, 2001. The middle class consensus and economic development. Journal of Economic Growth, 6(4), 317–335.

Ferreira, FHG & Gignoux, J, 2011. The measurement of inequality of opportunity: Theory and an application to Latin America. Review of Income and Wealth 57(4), 622–57.

Ferreira, FHG, Gignoux, J & Aran, M, 2011. The measurement of inequality of opportunity with imperfect data: The case of Turkey. Journal of Economic Inequality 9, 651–80.

Gornick, J & Jäntti, M (Eds), 2013. Income inequality: Economic disparities and the middle class in affluent countries. Stanford University Press, Stanford.

Heckman, J J, 2000. Policies to foster human capital. Research in Economics 54(1), 3–56.

Heckman, JJ, 2006. Skill formation and the economics of investing in disadvantaged children. Science 312(5782), 1900–2.

Hertova, D, López-Calva, LF & Ortiz-Juarez, E, 2010. Bigger ... but stronger? The middle class in Chile and Mexico in the last decade. Research for Public Policy, Inclusive Development, ID-02-2010, RBLAC-UNDP, New York.

Hoogeveen, JG & Özler, B, 2006. Poverty and inequality in post-apartheid South Africa: 1995–2000. In Bhorat, H & Kanbur, R (Eds.), Poverty and policy in post-apartheid South Africa. HSRC, Cape Town.

Kingdon, G & Knight, J, 2001. What have we learnt about unemployment from microdatasets in South Africa? Social Dynamics 27(1), 79–95.

Lam, D, Ardington, C, Branson, N, Goostrey, K & Leibbrandt, M, 2010. Credit constraints and the racial gap in post-secondary education in South Africa. Presented at annual meeting of Population Association of America, 15–17 April, Dallas, TX.

Lam, D, Ardington, C & Leibbrandt, M, 2011. Schooling as a lottery: Racial differences in school advancement in urban South Africa. Journal of Development Economics 95(2), 121–36.

Leibbrandt, M, Woolard, I, Finn, A & Argent, J, 2010. Trends in South African income distribution and poverty since the fall of apartheid. OECD Social, Employment and Migration Working Papers 101, OECD, Paris.

Mugero, C & Hoque, A, 2001. Review of Cholera Epidemic in South Africa, with Focus on KwaZulu-Natal Province. Provincial DoH, KwaZulu-Natal.

Özler, B, 2007. Not separate, not equal: Poverty and inequality in post-apartheid South Africa. Economic Development and Cultural Change 55(3), 487–529.

Paes de Barros, R, Ferreira, FHG, Molinas Vega, JR & Saavedra Chanduvi, J, 2009. Measuring inequality of opportunities in Latin America and the Carribean. World Bank, Washington, DC.

Piraino, P, 2012. Inequality of opportunity and intergenerational mobility in South Africa. Paper presented at the 2nd World Bank Conference on Equity, 27–28 June, Washington, DC.

Rawls, J, 1971. A theory of justice. Harvard University Press, Cambridge, MA.

Roemer, JE, 1998. Equality of opportunity. Harvard University Press, Cambridge, MA.

Seekings, J & Nattrass N, 2005. Class, race and inequality in South Africa. Yale University Press New, Haven, CT.

Sen, A, 1980. Equality of what? In McMurrin, S (Ed.), Tanner lectures on human values. Cambridge University Press, Cambridge.

Van der Berg, S, 2007. Apartheid's enduring legacy: Inequalities in education. Journal for African Economies 16(5), 849–80.

Van der Berg, S, 2009. The persistence of inequalities in education. In Aron, J, Kahn, B & Kingdon, G (Eds), South African economic policy under democracy. Oxford University Press, Oxford, pp. 327–54.

Van Der Berg, S, 2011. Current poverty and income distribution in the context of South African history. Economic History of Developing Regions 26(1), 120–40.

Visagie, J & Posel, D, 2013. A reconsideration of what and who is middle class in South Africa. Development Southern Africa 30(2), 149–67.

Woolard, I & Klasen, S, 2005. Determinants of income mobility and household poverty dynamics in South Africa. Journal of Development Studies 41(5), 865–97.

World Bank, 2010. How far are we from ensuring opportunities for all? The human opportunity index. The World Bank, Washington, DC.

World Bank, 2012. South Africa economic update: Focus on inequality of opportunity. The World Bank Group Africa region poverty reduction and economic management, (3). The World Bank, Washington, DC.

Yu, D, 2010. Poverty and inequality trends in South Africa using different survey data. Working Papers 04/2010, Stellenbosch University, Department of Economics.

Appendix A

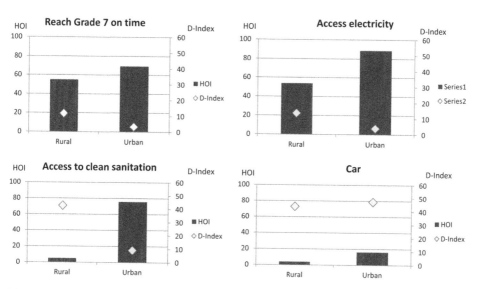

Figure A1: HOI for access to selected key basic goods for children age 10 to 14
Source: NIDS 2008.

Table A1: Logit school regression

Variable	(1) Age 10 to 14, Grade 7	(2) Age 19 to 20, matric
Mother lives in the household	0.0289	0.0348
	(0.0258)	(0.0517)
Mother's years of education	−0.0220**	−0.0370**
	(0.00894)	(0.0164)
Mother's years of education2	0.00326***	0.00409***
	(0.000721)	(0.00126)
Father lives in the household	0.0571**	0.0607
	(0.0283)	(0.0619)
Father's years of education	0.000249	0.0124**
	(0.00271)	(0.00502)
Log of per-capita household income	0.0294**	0.0495*
	(0.0140)	(0.0292)
Household has female head	0.00623	0.116**
	(0.0235)	(0.0531)
Number of children in household	−0.0143**	−0.0225
	(0.00608)	(0.0138)
Coloured	0.113***	−0.152**
	(0.0396)	(0.0606)
Indian	0.243***	−0.251***
	(0.0234)	(0.0451)
White	0.0159	−0.0154
	(0.0992)	(0.108)
Female	0.190***	0.144***
	(0.0205)	(0.0446)
Tribal	0.0342	−0.102
	(0.0407)	(0.0879)
Urban formal	0.0573	0.138
	(0.0435)	(0.0995)
Urban informal	0.0514	0.0847
	(0.0515)	(0.130)
Constant	−0.0199***	−0.00554***
	(0.6314)	(0.5087)
Observations	3305	1157

Note: Not listed: dummies for mother and father education if missing and provinces. Robust standard errors in parentheses. ***$p < 0.01$, **$p < 0.05$, *$p < 0.1$. Note that this is a logit model but marginal effects are reported.

Table A2: Panel regression

Variable	Ordinary least squares	
	(1) **Reach matric**	**(2)** **Get into tertiary education**
Mother's years of education	0.00712	0.00779**
	(0.00474)	(0.00376)
Father's years of education	0.00699	0.00404
	(0.00485)	(0.00366)
Log of per-capita household income	0.0435**	0.0732***
	(0.0201)	(0.0165)
Household has female head	0.0242	−0.0313
	(0.0433)	(0.0334)
Number of children in household	−0.0103	−0.00527
	(0.00946)	(0.00602)
Coloured	0.0495	−0.144**
	(0.0880)	(0.0689)
Indian	0.141**	0.0629
	(0.0717)	(0.236)
White	−0.174	−0.172
	(0.137)	(0.124)
Female	0.0273	−0.0292
	(0.0345)	(0.0277)
Tribal	−0.0853	0.000384
	(0.0789)	(0.0491)
Urban formal	0.00447	0.124**
	(0.0837)	(0.0591)
Urban informal	0.0762	0.146*
	(0.0999)	(0.0748)
Constant	0.0630	−0.512***
	(0.187)	(0.138)
Observations	1 467	1 467
R-squared	0.076	0.131

Note: Based on NIDS 2008 and 2010. Not listed: dummies for mother and father education if missing and province dummies. Robust standard errors in parentheses. ***$p < 0.01$, **$p < 0.05$, *$p < 0.1$.

Rethinking Bundy: Land and the black middle class – accumulation beyond the peasantry

Nkululeko Mabandla[1]

Based on an assessment of historical data on the black middle class in Mthatha, this article argues that South Africa's black middle class has considerable time depth. It originated in Bundy's 'peasantry', when African farmers started producing for the market and used their surpluses to educate their children. After being educated, these children continued to accumulate land for farming. Income from the land supplemented their salaries, which allowed them to further the education of their own children and accumulate additional land and, thus, wealth. Hence the black middle class in South Africa is arguably not a post-1994 phenomenon, but is rather the result of intra-generational transmission dating back to the mid-nineteenth century.

1. Introduction

As South Africa marks 20 years of liberation from apartheid, this historic process is not only cause for celebration but also, importantly, for reflection on the changes that have occurred since 1994. One topic that has drawn the attention of scholars is the question of the black middle class. It is often argued that this class only emerged (but then grew rapidly) in the post-apartheid period (Rivero et al., 2003; Udjo, 2008). Before 1994, colonialism and apartheid had imposed numerous constraints on Africans, including violent dispossession and a battery of laws that severely curtailed their ability to accumulate wealth and form a stable middle class. The liberal view holds that the development of a middle class is critical to social stability, as it mediates between rich and poor (see, for example, Lipset, 1968; Huntington, 1992). Against this backdrop, reflection on the history of South Africa's black middle class is particularly appropriate in the 20th year of democracy.

A central question is how long South Africa's black middle class has been in existence. Is it a recent, post-apartheid development or do class continuities date back much further? In considering this question, this article moves beyond the common conceptualisation of the black middle class as based on income and occupation (Crankshaw, 1997; Seekings & Nattrass, 2006), and instead examines a combination of occupation and property (land). For the black middle class in South Africa, wealth has historically meant land ownership; this will therefore be the focus of the discussion. The notion of wealth is essential for the conceptualisation of class in general. In societies marked by racial inequality, the structuring effects of accumulated wealth are often overlooked in favour of an almost exclusive emphasis on the removal of racial barriers to occupational mobility (on the racial distribution of wealth in the United States, see Oliver & Shapiro, 2006).

---

[1]Research Officer, Centre for African Studies, University of Cape Town, Rondebosch, Cape Town 7701, South Africa.

Drawing on a historical case study of class formation in Mthatha in the Eastern Cape, this article argues that South Africa's black middle class is a continuation of the successful black 'peasantry' described by Bundy (1988). Following Newman (1993:156), the article demonstrates the importance of historical continuities in considering the class question, involving 'family trajectory' as well as a 'shared identity' encompassing the whole family. The article assesses three generations of the Mthatha black middle class that emerged at the turn of the twentieth century, and shows intergenerational continuities in accumulation and cultural transmission that persist to the present day.

The paper is organised as follows: first, the historical context of middle-class formation in South Africa and its links to the nineteenth-century 'peasantry' are explored. The next section traces the development of the first generation of Mthatha's black middle class at the turn of the last century, focusing on the accumulation of land. A discussion of the development of the second generation of the middle class during the apartheid era follows, emphasising the formation of an ideological community and the central role of the women in class reproduction. The penultimate section reviews the development of the third generation during the former Transkei's 'independence' (1976–94) and until 2010. It assesses real-estate entrepreneurship in the context of agricultural decline and shows the continued relevance of land for this class. The conclusion links the historical discussion to contemporary debates. The argument presented in this paper is based on oral history interviews of about 30 descendants of Mthatha's black middle class, which were conducted in 2010, as well as on archival data (for a detailed discussion, see Mabandla, 2013).

2. Of peasants and middle classes – Bundy revisited

The period following the extension of colonial rule in the Cape is noted not only for rampant dispossession but also for growing differentiation within African society. Bundy (1988) highlights the influential role of the mission stations in instilling capitalist social and economic norms among their black converts. Missionary engagement on the Cape's Eastern Frontier differed markedly from the belligerent approach towards independent African polities beyond the colony's eastern border. It was influenced by the emerging middle-class culture in metropolitan Britain, which emphasised a common humanity and held that freedom from repressive laws was the basis for spiritual and material progress. On the Cape's Eastern Frontier, British humanitarianism advocated that extending Christianity and British civilisation to local populations was far more beneficial to the colonial order than military conquest and suppression (Lester, 2001). This set the stage for the development of the nineteenth-century black 'peasantry' (Bundy, 1988), and towards the end of that century a black middle class.

A careful reading of Bundy's work shows the complex intersection of cultural and economic institutions, simultaneously preventing and enabling access to land by black people. Colonial conquest disrupted the stability of the precolonial mode of production through the dispossession of African land. This was followed by colonial land reforms, which saw large tracts of productive land privatised and concentrated in European hands. On the other hand, the mission stations facilitated limited land accumulation by Christian converts throughout the 1830s. By the 1840s, the Cape colonial government had issued title deeds to about 70 000 acres of land to mission stations (Bundy, 1988). Colonial land reforms had both economic and political implications for African social formations. First, the dispossessions disrupted traditional landholding patterns and production structures; second, the subsequent land redistributions disrupted the chiefs' power to allocate land. Land-poor adherents could now turn to missionaries (and conversion to Christianity) as an alternate option for accessing land, thus bypassing traditional structures.

The combination of cultural and economic factors meant that social mobility was a question not only of access to land but also increasingly of education. Mission stations contributed to education-based stratification by training converts in new agricultural techniques, literacy and numeracy in order to facilitate economic transactions. This improved the productive capacity of African farmers and contributed to the formation of a class of agricultural producers who could participate fully in the emerging capitalist agrarian economy. Increasing demand for commodity production, the imposition of colonial taxes and the general penetration of the rural economy by consumer goods compelled them to expand beyond subsistence farming to produce a surplus for the market. By adopting a mixed model of subsistence and commercial agriculture, they laid the foundation for the development of a small black middle class in the Eastern Cape. They reinvested their profits in equipment, livestock and crops, as well as in the education of their children, who became clerks, teachers, priests and even interpreters. This education represented a qualitative shift from the functional training their parents had received. The children had marketable skills – in the sense of Weber (1968) – that could be exchanged on the emerging capitalist labour markets, and they reinvested their salaries from these occupations in agriculture.

In addition to improving the productive capacity of local farmers, the missionaries established stores that sold clothing, household items and agricultural implements. Mission stations thus contributed to both the transformation of the mode of production and the establishment of new forms of consumption. They encouraged new ways of living, including the move towards square houses (instead of the traditional round houses or rondavels). The new type of house was symbolic of new patterns of consumption and tied African societies firmly to the British colonial economy. As elaborated in the missionary journal, the *Kaffir Express*: 'With a proper house, then comes a table, then chairs, a clean cloth, paper or whitewash for the walls, wife and daughters dressed in calico prints, and so forth' (cited in Bundy, 1988:37). The logic of consumerism embedded in the design of living space was again evident a century later, with the introduction of 'matchbox' houses in African townships. Upon receiving a house, a first-generation urban resident is reported to have remarked: 'This kind of house whispers to you that it needs more furniture' (Bonner, 1995 cited in Seekings & Nattrass, 2006:103).

In summary, mission stations played a major role in colonial social transformation. Access to land, education, a combination of subsistence and commercial agriculture, and lifestyle and consumption changes contributed to social stratification within African society and led to the rise of new, post-traditional classes, the 'peasantry' (Bundy, 1988) and the black middle class.

By and large, African society adapted to these changes. The training they obtained gave black agricultural producers a competitive edge and they were soon able to increase their output significantly. However, the transition into agrarian capitalism also exacerbated inequalities between rural households. Higher demand for food production meant that those who could afford to purchase ploughs, wagons and other equipment took full advantage of market opportunities and competed successfully with white farmers.

However, this progress was soon crushed through the adoption of the Natives Land Act of 1913. Bowing to pressure from the commercial farming and mining industries to address labour shortages, the state introduced a singularly controversial piece of legislation. Under this Act, Africans were forbidden from accessing and owning land outside the reserves. Black sharecroppers and labour tenants, who had worked the land as independent producers in areas now reserved for white occupation, were most affected. They were forcefully removed to the reserves created by the Glen Grey Act of 1894, which amounted to only around 7% of the country's land area.

The 1913 Act caused massive dispossession of black farmers. According to Bundy (1988), it led to the 'fall' of the 'peasantry', as overwhelming numbers of farmers were forced into wage employment for survival (also Mafeje, 1988; Plaatje, 1995). However, the grand narrative of proletarianisation cannot adequately explain the persistence of landownership within certain sections of black society. The literature is relatively quiet on the fate of the descendants of the early commercial farmers. Did all of them enter the ranks of the proletariat or other forms of wage labour? Did access to land disappear entirely? As the following case study of Mthatha's black middle class demonstrates, land accumulation for these descendants continued, in an urban context, beyond the 'fall'.

3. Elisha Mda and the first generation of Mthatha's black middle class

The story of Elisha Mda, one of the early black landowners in Mthatha, is important for understanding the colonial history of the black middle class and the interplay between landownership and the opportunities afforded by mission education. Described in Mabandla (2013), this story sheds light on the lives of African landowners following the restrictions on land accumulation in the Cape and demonstrates the continuity of the middle class beyond Bundy's 'fall of the peasantry'. In this article, Mda's life history illustrates the linkages between the 'peasantry' and the first generation of the black middle class.

Elisha Mda's date of birth is unknown but is generally associated with the great drought of 1860. It could well have been earlier, as he had qualified as a teacher by the late 1870s. His family had settled in Dutywa after their expulsion from the mouth of the Buffalo River following the Nongqawuse cattle-killing disaster (1856–58). Mda grew up in a traditional context; his parents were peasant farmers similar to those described by Bundy. They invested in his education and sent him to live with his aunt, who stayed near a missionary school. His grandson, Mda Mda, explains:

> We were red people from Ndlambe's territory. Elisha went to stay with his aunt in Tsomo with the amaZizi clan. A missionary school had been established there by the whites. That land was under white rule. So, he attended school. And because the boy was clever, he was well liked by the teachers and missionaries. (M Mda, interview, Mthatha, 9 July 2010)

After completing his education at Lovedale, Mda became one of the first teachers at a newly opened school in Ntshatshongo (Fort Malan) in the 1870s. He also managed to acquire land, by saving part of his salary and investing his savings in wagon transport. He then combined teaching with transport riding until he had accumulated enough money to buy land. Bundy (1988:77) suggests that this was common practice. Many 'peasants' engaged in transport riding once their crops were harvested, taking their produce to the markets in Queenstown and King William's Town and carrying goods for local traders on the return journey to Fingoland.[2] Transport riding, according to Bundy, was a means of earning money to buy or hire land, and many did this on a full-time basis. Mda used the proceeds from transport riding to buy land in British Kaffraria.[3] However, like many others, he lost his land after being suspected of aiding and abetting enemies of the British Crown in the last frontier war, the War of Ngcayichibi (1877–78). It is not clear whether he or any of the other landowners received compensation, but he reinvented himself as a landowner in Mthatha soon after, at the turn of the twentieth century.

The first generation of Mthatha's middle class illustrates the historical complexity of black landownership and agricultural production. The early adoption of education and the combination of farming and other occupations render Bundy's term 'peasant' inappropriate for people like Elisha Mda. It also shows that the 'many Africans' engaged in transport riding (Bundy, 1988:77) were in fact a highly differentiated group. The few who could afford land were drawn mostly from the limited ranks of professionals (Peires, 1989).

The emergence of a substantial and well-defined first generation of the black middle class in Mthatha is closely linked to the Umtata Water Scheme of 1906, when municipal land had to be auctioned to finance the building of the then Kambi (now Mthatha) Dam. After many attempts to restrict land sales to whites, black farmers were eventually allowed to take part in the land auction of Plot E, later known as Ncambedlana, in 1908. This laid the foundation for land accumulation by Mthatha's black middle class, a process discussed at length in Mabandla (2013).

The town of Mthatha was an important commercial and administrative centre. Its well-developed railway network linked the Transkei hinterland with the industrial centres of the Republic. The town attracted both labourers en route to cities (such as the industrial metropolis of Johannesburg) and the aspiring middle class who held professional positions in the colonial civil service. Many members of the first-generation middle class came from the ranks of the latter group. Apart from Elisha Mda, noted members of this generation included AC Zibi, Luke Yako, David Noah and Tennyson M Makiwane (Redding, 1992). Most were descendants of Bundy's 'peasantry', who had obtained an education and established themselves in white-collar occupations in town, using their salaries to accumulate land. Bundy's 'peasantry' could thus be described as 'Generation Zero', who used profits from agriculture to educate their children. They, in turn, became 'Generation One' – Elisha Mda and others like him who had professional jobs and accumulated land in places such as Mthatha. This continuity with the activities of their 'peasant' parents is not captured by the 'fall' thesis.

The contrasting backgrounds of TM Makiwane and Elisha Mda illustrate the complex history of this class, in which some were closer to their traditional/peasant roots, while others had a longer family history of education. TM Makiwane was the son of Elijah Makiwane, a mid-nineteenth-century leader of the church and the African education movement. Hence, unlike Mda, Makiwane did not come directly from the 'peasantry'; he was essentially second generation, since his father had been educated.

---

[2] Area between the Kei and Mbashe Rivers.
[3] Encompassing the present-day districts of East London and King William's Town.

A graduate of Lovedale College, he was a teacher until 1910, when he bought land in Mthatha and left teaching to take up a senior clerical position at the Transkeian Territories General Council – the *Bhunga* (caucus, isiXhosa).[4] Makiwane was active in local politics and was an elected member of many civic organisations (Redding, 1992; for more details on his life, see Mabandla, 2013). His opposition to white domination at the *Bhunga* (M Mda, interview, Mthatha, 9 July 2010) and his civic spirit are celebrated in the poem *Izibongo zika Gambu* (The Praises of Gambu, isiXhosa) by St J Page Yako.[5] Makiwane was also the first editor of the bimonthly English–isiXhosa agricultural journal, *Umcebesi Womlimi Nomfuyi/Agricultural and Pastoral Guide* (founded in 1925 and published by the Agriculture Department of the United Transkeian Territories General Council; see Switzer & Switzer, 1979) as well as of the local weekly *Umthunywa* (The Messenger, isiXhosa). While maintaining his clerical position, he was also a highly successful farmer (M Makiwane, personal correspondence, 1 August 2010). His salary and agricultural profits enabled him to educate not only his children but also his wife Virginia, who established a career as a social worker. In 1943 he bought six more plots at a cost of £800, at a time when an experienced African clerk earned only around £160 per year (Redding, 1993).

The success of the first generation of Mthatha's black middle class lay in their unique combination of education, employment and land ownership. A mixed model of subsistence and commercial farming, combined with salaries from professional jobs, formed the basic mode of reproduction for this class. They continued to farm as their parents did, raising both crops and livestock. In addition to their own land, they had access to the municipal commonage for grazing their livestock. Such access was crucial to their success, and the decline of farming in later generations is partly explained by housing development on the Mthatha commonage.

Mthatha's first-generation black middle class shows that there was little discontinuity in their lives from being successful 'peasants' in the rural context to combining professional jobs with landownership and cultivation. Thus, in contrast to Bundy's 'fall', there was continuation and even an ongoing 'rise'.

4. The second generation, 1950s to 1976

Between 1950 and 1976 the ruling National Party severely tightened segregation laws and implemented various apartheid policies. Among these was the Group Areas Act of 1950, which demarcated separate residential and business areas for blacks and whites. In the traumatic forced removals that followed, large numbers of African people were herded into locations far from their places of work and business. Yet while people were forcibly being removed from areas such as Sophiatown and District Six, the second generation of the Mthatha black middle class continued to acquire land and urban property.

Many scholars hold that the development of the black middle class in this period followed an occupational trajectory, resulting from black urbanisation in the 1940s–60 s (Kuper, 1965; Crankshaw, 1997; Seekings & Nattrass, 2006). Even in the Bantustan context, the one place where blacks could own land, research has largely focused on the role of the Bantustan state in enabling class formation (Southall, 1982;

---

[4] This was a form of local government based on district councils of elected black members under the chairmanship of white magistrates.
[5] *Umcebesi Womlimi Nomfuyi/Agricultural and Pastoral Guide*, May 1950.

also Josana, 1989), while the role of land has largely been overlooked. One of the few studies addressing land (Redding, 1993) also argues that commercial farming in Mthatha, which had protected the middle class from sliding into migrant labour, had declined by 1950. However, the part of the middle class that combined professional employment and landownership continued to exist after the 1950s.

The ongoing accumulation of urban land by this middle class was unusual. Towns like Mthatha, with significant European commercial interests, were not part of the reserves. They were regarded as 'white spots' and were thus subject to the same segregation legislation that had transformed urban communities elsewhere. Yet Mthatha's black middle class was adept at using local politics to protect its land rights. Tensions around black landownership in Mthatha had surfaced as the governing National Party sought to enforce segregation. Local Afrikaners, emboldened by the party's rise, had petitioned the government about what they saw as the deplorable 'mixing of the races' in Ncambedlana and requested it to act against black ownership in line with the Group Areas Act (Redding, 1992). Mthatha's black middle class responded with their own campaign, lobbying the government to declare Ncambedlana a black area, 'thrown open' for African ownership. This was, however, declined by the Minister of Native Affairs, HF Verwoerd (who became Prime Minister in 1958). Verwoerd outlined government policy as follows:

> Umtata is at present the centre of all governmental activities in the Transkeian Territories and is also probably the largest and most important commercial centre. As a result it has a large European population, and many of the Europeans have large vested interests involving a considerable amount of capital. For the moment this centre accordingly is regarded as European in character and in general the policy of the government in regard to it must be identical with that pertaining elsewhere throughout the Union. (MAR, 1951)

The campaign by Mthatha's black middle class was a calculated move: they had reasoned that excising Ncambedlana from the municipality to allow for black occupation would have been a highly complicated and controversial undertaking and that the government would have preferred to maintain the status quo. A number of white people owned properties in the area and would have had to be compensated. Most whites in the Transkei reserve, including the Umtata Municipal Council, were aligned to the opposition United Party, which had been defeated by the National Party in the 1948 elections (Southall, 1982:149), and they would probably have resisted such a move. The gamble paid off: black landowners were able to retain their properties, as Verwoerd confirmed the distinctive challenge this presented:

> I realise that the position at Ncambedlana is unique in that the area borders on the Native Reserves and that the general authority permitting the acquisition of freehold title to the lots in that area was granted with the full consent of the local authority. I have, therefore, agreed that the position there should remain unchanged, although I would prefer to see Ncambedlana excised from the municipal area. (MAR, 1951)

This decision led to two simultaneous processes: National Party adherents exerted pressure on black residential and trading rights in the central business district in order to monopolise these for whites, while the black middle class bought out almost all of the white-owned properties in Ncambedlana. As a result, the second generation of the black middle class was established before the full implementation of the Bantustan strategy, and was firmly entrenched by the time Transkei was accorded self-governing status in 1963.

While the first and second generations had roughly the same professional backgrounds, the latter attended university (Fort Hare) rather than college. This educational mobility gave them access to a more diversified range of occupations (e.g. teachers, lawyers, doctors and entrepreneurs), better-paid jobs and a more secure income base. It also conferred a status that transcended the local setting, since they attended university with many of the major nationalist leaders. According to Aunt Laura Mpahlwa, who settled in Ncambedlana in the 1960s, the black middle class of the time was a highly status-conscious social group, who would occasionally flaunt their university education to distinguish themselves from others:

> The majority of them were teachers, eh, the majority of them had degrees, studied at Fort Hare and Lovedale, and we used to boast about those things. That if you didn't study in Lovedale or Fort Hare then, you were really a nobody. You were a nobody. But in their simple way, they were people who were accommodative; they did not really look down upon those who really were not able to reach the highest educational standards.
> (L Mpahlwa, interview, Mthatha, 20 July 2010)

Marriage patterns also showed significant consolidation: members of this social group had generally attended the same missionary schools as their spouses and had married within their class. Social and cultural capital in the sense of Bourdieu (1984) circulated within and around this class. Recruitment to jobs, especially at prestigious black schools such as St John's College, utilised the social networks established at Fort Hare. Information about land sales was also circulated in these networks. Ultimately, unlike the first generation, which had lived elsewhere and used the area mainly for farming, a settled middle-class community with strong social bonds developed in Ncambedlana. They built family homes, schools, churches and medical facilities. In the cultural realm, aspirational middle-class behaviour included the organisation of choral eisteddfods,[6] with competitors coming from as far afield as Kimberley.

The sense of community among the second generation was based on a commitment to middle-class values and a strong communal identity. Nomonde Bam, whose father-in-law (Ngubethole Bam) was a member of the second generation, describes these close bonds of solidarity and support in times of joy and sorrow: 'whether it's a wedding or funeral or whatever', people were there for each other. All were involved in planning and logistical support, whether through 'prayers' or more materially through food, for example, 'because there would be visitors throughout the week ... one didn't wait to be asked for support' (N Bam, interview, Mthatha, 19 July 2010).

As with the first generation, the second generation continued to combine professional employment and agriculture. However, gender discrimination meant that married women in this generation found their career prospects considerably diminished. In the teaching profession especially, only single women and men were considered for permanent positions (Kotecha, 1994; also Van den Heever, 1975). As a result, married women soon found themselves forced into the domestic sphere. These women, many

---

[6]A Welsh festival of literature, music and performance dating back to the twelfth century.

of whom had equivalent qualifications to their husbands, were not only central to domestic production but were also responsible for the management and control of agricultural production, the market sale of crops, and the contracting and remuneration of labour. This included the organisation and management of hired labour and the *amalima* (cooperative labour groups, isiXhosa; McAllister, 2004), the two main forms of agricultural labour used.

Helen Bradford's (2000) critique of Bundy's work on the 'peasantry' focuses on the silencing of the women's role. According to her, Bundy's 'peasants' had displaced women from their traditional domain – agriculture – once an agricultural market had developed. This line of argument, however, obscures the underlying social relations of power and control. Lewis (1984) highlighted the gendered nature of the precolonial mode of production, which was divided into agriculture as the women's sphere and cattle rearing as a male domain. This provided cultural justification for women's dependence on men, since cattle outranked cultivation in the food cycle. While the central role of these black middle-class women in agriculture might suggest a gendered recovery of this sphere, livestock remained a male preserve among the second generation of Mthatha's middle class.

In addition to these roles, the women also played a key role in developing and reproducing the attributes of their middle-class status. They opened informal crèches, where local children were schooled before enrolling in primary schools, and instilled and supervised standards of behaviour and propriety (also Wilson & Mafeje, 1963). While the men went to work, they remained in the community, where they could discourage neighbourhood children from anti-social behaviour or give horticultural 'advice' to those whose houses appeared unkempt. They formed or joined typical middle-class associations, such as the Young Women's Christian Association and the *Zenzele* (do it yourself, isiXhosa). Their everyday practices and activities were thus essential to the middle-class 'habitus' (Bourdieu, 1984). Middle-classness, in other words, is not just a question of ownership, occupation or income, but also of values and policed behaviour. The role of the women challenges the dichotomy that is often assumed by dividing social reproduction into a dynamic-productive-male realm and a passive-domestic-female one. While being central to agricultural production, these women were equally central to the reproduction of social relations. The socialisation of children along middle-class values occurred inter-generationally, and these social relations changed and were reproduced intra-generationally as the children grew up. This is clearly demonstrated by the influence of the women on the educational success of their children. As explained by Aunt Laura Mpahlwa:

> *Abantwana* [the children, isiXhosa] got a very good background of English, what then used to be called Royal Readers. The people who were teachers were Royal Readers so, the children of Ncambedlana knew English like anything and they spoke good English. They wrote good English and were getting very good marks. (L Mpahlwa, interview, Mthatha, 20 July 2010)

Just as with the first generation, the next generation continued with the mixed model of agriculture and reinvested their profits in their children's education. As one of their descendants, Loyiso Mpumlwana, puts it:

> ... there was a lot of this subsistence farming and commercial farming. People would *consume* and also *sell* ... *people managed to put their kids to school. We got our education on the basis of commercial farming.* We

had a donkey cart at home and we sold. We would load it and sell here in town. (L Mpumlwana, interview, Mthatha, 26 July 2010; emphasis added)

One reason for the farming continuities within this class was that black incomes from professional occupations, especially in the civil service, remained notoriously low. As in the first generation, the combination of subsistence and commercial agriculture allowed black middle-class families to thrive despite poor salaries. Thus, the 'rise' continued: not only did the peasantry not fall, but the middle class that emerged out of the peasantry also continued to flourish after 1913 and even during apartheid.

5. The third generation, 1976–2010

The proclamation of the Transkei as an 'independent homeland' broadened the occupational base of the middle class through the opportunities offered by the Bantustan state structure. Such new opportunities opened up in administration and management (and even business) as the Africanisation policies of the Transkei regime sought to replace white officials with black ones. It is no surprise that Mthatha's black middle class – drawing on generations of education – was well placed to exploit the new environment. (The same could be said about black advancement in the democratic era.)

In this process, the third generation of the Mthatha middle class was transformed along occupational lines. While the bulk of the first and second generations could be classified in the lower to middle income or education categories (following Crankshaw, 1997; Seekings & Nattrass, 2006), the next generation was more professionalised. The number of university graduates increased, and there was a growing diversity of occupation, including medicine, law, management and business, for example. This allowed people to improve and broaden their income base even further. As one member of the third generation, Sembie Danana, puts it: 'very few went into teaching' (S Danana, interview, Mthatha, 18 July 2010). Thus, over time, the middle class had been transformed from one that combined employment and farming in the first and second generations to one that that was educationally distinguished, working mainly in prestigious professional positions. (Members of the middle class went on to find well-paid employment across the country, and after democratisation many rose to even higher positions, supported by the legislative framework of employment equity and black economic empowerment.)

These new opportunities continued the non-discriminatory practices of the 'self-government' and 'independence' eras, which had also allowed women such as Aunt Laura Mpahlwa and Mrs MMM Raziya to come to prominence in the business and public spheres. For example, Mrs Raziya served for many years as the only female councillor in the Umtata City Council (Umtata City Council, 1982). In politics too, women carved out larger roles; for example, Stella Sigcau (later the Minister of Public Enterprise in the Mandela administration) was appointed Prime Minister of the Transkei in 1987.

These socio-political changes affected the agricultural activities of the middle class. As they had aged, the second generation had become increasingly unable to continue the demanding work of agriculture. For the third generation, however, the situation was different. The curriculum of the missionary schools, taught to their parents and grandparents, focused on unity of mind, body and spirit. Ploughing the land was regarded as important for the students' sustenance, as well as their health and mental well-being. They were trained in agricultural skills, so that 'they might be afterwards able to instruct their countrymen in the art of cultivating their own soil' (Shepherd, 1940:90). In contrast, for the third generation, training in industrial and agricultural skills had been removed from the academic syllabus following the centralisation of education under the Department of Education and Training. Such skills were now taught at special vocational schools (Christie & Collins, 1984). Thus the third generation lacked the enthusiasm and skill of their forebears. In addition, agriculture had gradually been 'stigmatised' as the last resort of those who could not obtain a university education. In the Transkei, in particular, the education system was geared mainly towards equipping people for urban white-collar professions (Ntsebeza, 2006).

Apart from the school syllabus, other factors leading to the decline of agriculture include the agricultural policies of the Bantustan regime, which favoured large-scale commercial agriculture over household agricultural production or the family farms of the Mthatha middle class. To encourage rapid urbanisation, housing development was undertaken on the commonage that had served as grazing lands for the first and second generations. The suburbs of Hillcrest, Northcrest and Hillcrest Extension were developed in the early 1970s, 1981 and 1987 respectively (Siyongwana, 1990). This effectively pushed cattle, which were needed for ploughing, out of the urban environment. Ecological factors, such as the 1980s drought, also contributed to the decline of agriculture. A final factor was women's growing professional mobility, which meant their withdrawal from household agricultural production as well as from community building. The opportunities that had opened for men and women of the third generation thus affected both agricultural production and the character of the neighbourhood. Out-migration to larger urban centres also negatively affected community cohesion and lowered the social capital of those who stayed behind (see, for example, Beatty et al., 2009). However, land remained important for the identity and well-being of the middle class.

In the democratic era, some of these factors continued to influence land-use patterns, as did labour shortages resulting from new opportunities for those who historically had provided agricultural labour. In addition, urban population pressures contributed to the conversion of farmland for rental accommodation. As in the rest of urban South Africa, Mthatha's population has increased significantly because people from the surrounding villages moved into town in search of employment opportunities, mainly in white-collar professions and social services (Makgetla, 2010). The population of Mthatha is reported to have increased from 72 000 in 1991 to 91 000 in 2001 (Siyongwana, 2005), while later estimates put the figure at around 150 000 (Harrison, 2010). This, coupled with the inadequate provision of low-cost housing during the 'independence' era (Siyongwana, 2005), has increased the demand for housing, which has in turn enhanced the position of landowners. There is differentiation among the landowners in this regard: those with more financial resources undertake large-scale suburban developments, while those with fewer resources invest in smaller projects. Intermediate housing developments involve the building of several rental units, and those with limited resources let individual rooms on their property. Real-estate entrepreneurship has become a way of life for the third generation, just as agriculture had been for the previous generations.

6. Conclusion

This article refutes conventional wisdom by demonstrating that the emergence of the black middle class predates democratisation. The early beginnings of this class were traced to Bundy's 'peasantry', which developed in the Cape during the nineteenth century. The article has pointed to important historical continuities in considering the class question. It has shown the relationship between land and education, which has been at the centre of social change within African society, and its continued role in defining family 'trajectories' and class 'identities'. This relationship allowed the development of what we call Generation Zero, Bundy's 'peasantry', and later that of Generation One, the early black middle class of Elisha Mda and others. His example, in particular, illustrates dispossession and ownership in the lifespan of one individual: land lost, land obtained, and land lost again. His education, the one thing that could not be taken from him, gave him a new chance for new land in Mthatha.

The symbiotic relationship between land and education continued to play a pivotal role in reproducing two more generations of the black middle class, the second and third generations. This article has highlighted the central role of women in wealth creation and cultural transmission, at a time when workplace discrimination (especially in the teaching profession) excluded many married women from the workforce. It has shown the ongoing importance of land to the middle class in a period of agricultural decline, as evidenced by the present-day real-estate entrepreneurship of the third generation.

The historical continuities, albeit nuanced, in accumulation, education, farming and agricultural practices demonstrate the strong links between the nineteenth-century 'peasantry' and the middle class that emerged in the Cape in the late nineteenth and early twentieth centuries. While the case of the Mthatha middle class cannot be generalised, this development may not be unique to the liberal conditions of the old Cape Colony, which initially allowed black landownership. Similar developments may well have been replicated elsewhere, especially in the context of the Bantustans, where Africans could own land after 1913.

Brandel-Syrier's (1971) study of the black middle class of Reeftown also shows that the success of this class in the urban environment depends on both education (and, hence, occupation) and the persistence of rural landownership. Methodologically, most current approaches to the black middle class focus on income and quantitative analysis, leading them to locate the emergence of this class firmly within the democratic era. They overlook Bourdieu's (1984) concept of 'habitus', which mediates the intra-generational reproduction of social relations. Thus, the question of land and wealth transmission has remained obscured by contemporary sociology's overwhelmingly urban focus and its fixation on occupations and incomes.

References

Beatty, C, Lawless, P, Pearson, S & Wilson, I, 2009. Residential Mobility and Outcome Change in Deprived Areas – Evidence from the New Deal for Communities Programme. Department for Communities and Local Development, London.

Bourdieu, P, 1984. The forms of capital. In Richardson, J (Ed.), Handbook of theory and research for the sociology of education. Greenwood, New York, pp. 241–58.

Bradford, H, 2000. Peasants, historians, and gender: A South African case study revisited, 1850–1886. History and Theory 39, 86–110.

Brandel-Syrier, M, 1971. Reeftown elite: A study of social mobility in a modern African community on the Reef. Routledge, London.

Bundy, C, 1988. The rise and fall of South African peasantry. David Phillips, Cape Town.

Christie, P & Collins, C, 1984. Bantu education: Apartheid ideology and labour reproduction. In Kallaway, P (Ed.), Apartheid and education. The education of Black South Africans. Ravan Press, Johannesburg, pp. 160–83.

Crankshaw, O, 1997. Race, space and the post-Fordist spatial order of Johannesburg. Urban Studies 45, 1692–711.

Harrison, K, 2010. Can Mthatha turn the corner? SA Delivery 18, 24–7.

Huntington, SP, 1992. Democracy's third wave. Journal of Democracy 2, 12–35.

Josana, X, 1989. The Transkeian middle class: Its political implications. Africa Perspective 1, 94–104.

Kotecha, P, 1994. The position of women teachers. Agenda 21, 21–35.

Kuper, L, 1965. An African bourgeoisie: Race, class and politics in South Africa. Yale University Press, New Haven, CT.

Lester, A, 2001. Imperial networks: Creating identities in nineteenth-century South Africa and Britain. Routledge, London.

Lewis, J, 1984. The rise and fall of a South African peasantry: A critique and reassessment. Journal of Southern African Studies 11(1), 1–24.

Lipset, SM, 1968. Stratification: Social class. In Sills, DL, International encyclopedia of the social science 15. Collier Macmillan, New York, pp. 296–316.

Mabandla, N, 2013. Lahla Ngubo: The continuities and discontinuities of a South African Black middle class. African Studies Centre, Leiden.

Mafeje, A, 1988. The agrarian question and food production in southern Africa. In Prah, KK (Ed.), Food Security Issues. Selected Proceedings of the Conference on Food Security Issues in Southern Africa, Maseru, 12–14 January 1987. Southern African Studies Series 4, Institute of Southern African Studies, National University of Lesotho.

Makgetla, N, 2010. Synthesis paper: South Africa. Development Bank of Southern Africa. Unpublished paper. http://www.rimisp.org/FCKeditor/UserFiles/File/documentos/docs/sitioindia/documentos/Paper-Country-Overview-South-Africa.pdf Accessed 12 March 2012.

MAR (Mthatha Archives Repository), 1951. Letter by HF Verwoerd (Minister of Native Affairs) to the Secretary, Ncambedlana Ratepayers Association, 15 September. MAR 4/24/13/2, file no. 4/10/2.

McAllister, P, 2004. Labour and beer in the Transkei, South Africa: Xhosa work parties in historical and contemporary perspective. Human Organization 63, 100–11.

Newman, KS, 1993. Declining fortunes: The withering of the American Dream. Basic Books, New York.

Ntsebeza, L, 2006. Democracy compromised: Chiefs and the politics of land in South Africa. HSRC Press, Cape Town.

Oliver, ML & Shapiro, TM, 2006. Black wealth, White wealth: A new perspective on racial inequality. 2nd edn. Routledge, London.

Peires, JB, 1989. The dead will arise: Nongqawuse and the great Xhosa cattle-killing movement of 1856–7. Ravan Press, Johannesburg.

Plaatje, ST, 1995. Native life in South Africa. Ravan Press, Randburg.

Redding, S, 1992. South African blacks in a small town setting: The ironies of control in Umtata, 1878–1955. The Canadian Journal of African Studies 26, 70–90.

Redding, S, 1993. Peasants and the creation of an African middle class in Umtata, 1880–1950. The International Journal of African Historical Studies 26, 513–39.

Rivero, CG, du Toit, P & Kotze, H, 2003. Tracking the development of the black middle class in democratic South Africa. Politeia 22, 6–29.

Seekings, J & Nattrass, N, 2006. Class, race, and inequality in South Africa. Yale University Press, New Haven, CT.

Shepherd, RHW, 1940. Lovedale South Africa: The story of a century, 1841–1941. Lovedale Press, Lovedale.

Siyongwana, PQ, 1990. Residential transformation in Umtata since 1976. Unpublished MA dissertation, University of Port Elizabeth.

Siyongwana, PQ, 2005. Transformation of residential planning in Umtata during the post-apartheid transition era. GeoJournal 64, 199–213.

Southall, R, 1982. South Africa's Transkei: The political economy of an 'independent' Bantustan. Heinemann, London.

Switzer, L & Switzer, D, 1979. The Black press in South Africa and Lesotho: A descriptive biographical guide to African, Coloured and Indian newspapers, newsletters and magazines, 1836–1976. G.K. Hall & Co, Boston, MA.

Udjo, EO, 2008. The demographics of the emerging black middle class in South Africa. Research Report 375, Bureau of Market Research, University of South Africa, Pretoria.

Umtata City Council, 1982. Umtata history, centennial anniversary publication. Government Printer, Pretoria.

Van den Heever, J, 1975. Some rights and wrongs. South African Outlook, July, p. 109.

Weber, M, 1968. Economy and society: An outline of interpretative sociology. Bedminster, New York.

Wilson, M & Mafeje, A, 1963. Langa: A study of social groups in a South African township. Oxford University Press, Cape Town.

What middle class? The shifting and dynamic nature of class position

Grace Khunou

Class categorisation should not only be informed by academic pursuits but by the lived experiences of those being categorised. A human or community-centred definition of class will illustrate the complexities of class experience and will thus present a dynamic conceptualisation. Through two life-history interviews of two black women from South Africa, this article illustrates that middle-classness for blacks during apartheid was marred with constant shifts related to the socio-economic and political impermanence of class position. Continuous negotiation driven by the need to be included in one's own community and the effects of being racially othered in interaction with whites and white spaces influences these shifts. In conclusion, the article argues that being middle class and black is heterogeneously experienced and thus complex.

1. Introduction

Most recently, South Africa (SA) has seen a growth in studies trying to understand the black middle class. These have been varied in their approaches and intentions. There have been those that are more retail oriented, populist and reductionist in their approach and conclusions. These have reduced the experience of the black middle class to an undifferentiated mass of conspicuous consumers, foregrounding the tradition of conceptualising class and general life experiences of black people as homogeneous and fixed. Other studies have been more theoretically grounded and progressive in their contribution to knowledge production and societal illuminations of experiences of class, meanings of class and the complexities of the language used to denote individual social class positioning (Phadi & Ceruti, 2011; Phadi & Manda, 2010; Krige, 2011b).

Since the end of apartheid there has been a move to understand the black middle class. This is so especially after the enactment of legislations to address past inequalities like affirmative action and the Black Economic Empowerment imperatives. With these changes SA saw an exponential growth in the black middle class. Given the various conceptions of what constitutes a black middle class it is difficult to reconcile studies on its size and constitution (Visagie, 2011). However, research evidence indicates a growth post 1994 (Krige, 2011a; Visagie, 2011). Although this change has been recorded as positive by many it has nevertheless been accompanied by growing inequalities, with SA's Gini coefficient 'increasing from 0.64 in 1995 to 0.72 in 2005' (Bhorat, Van Der Westhuizen and Jacobs 2009:12). These increasing levels of inequality have been more intra-racial and illustrate a shift from the historical inter-racial inequalities known for apartheid SA (Seekings & Nattrass, 2002; Leibbrant et al., 2010). This should not, however, be read to mean that racial inequality has been eroded (Keswell, 2010; Gumede, 2011; Leibbrandt, Wegner and Arden, 2011), but that growing numbers of black people are in the middle class with a few more in the upper class (Seekings & Nattrass, 2002; Leibbrant et al., 2010).

This article is based on life-history interviews of two women who reluctantly self-identify as middle class. Given the relatively high educational attainment of both participants,[1] their understanding of class was an invocation of theoretical understandings of the concept lived experiences of racial segregation and renegotiation of social positioning. Thus their general conception of their social position as middle class was critically derived from their shifting experiences based on where they were and who was present (Lacy 2007). Being black in apartheid SA meant their being middle class came with constant complex negotiations of boundaries with community members that were not middle class and spaces that were middle class but white, thus raising racial dynamics not experienced at home. Although they experienced some changes in their class position in contemporary SA that are not discussed in this article, their experiences were marred by constant shifts and everyday negotiations. These complexities were a result of the socio-economic and political impermanence of their middle-class position, competing social inclusion needs and also the constantly shifting membership to this class.

This article provides a discussion of how these two women experienced class and how the flux nature of the position provides significant pointers for a critical re-examining of how the black middle class is lived and experienced from a subjective point of view. The next section provides the theoretical discussion of the concept class and middle class. The discussion that follows then presents a brief explanation of the methodology, which is followed by a detailed discussion of the findings from the two life histories. Finally, a brief conclusion and summary is provided.

2. Class: Some theoretical discussions

The notion of conspicuous consumption that is linked to the black middle class erroneously suggests that blacks as members of this class consume for the sake of consumption. Conspicuous consumption has been defined as purchasing a product not for its utility but for displaying wealth and purchasing power, where the 'price becomes the only factor of any significance to him or her' (Mason, 2007:26). On the contrary, Krige's (2012) reading of Soweto suggests that on 'a closer look at the longer histories of social mobility, social distinction and consumption provides us with a more complex and nuanced reading of the possible meanings of consumption'. Krige (2012) then suggests that consumption linked to house building and renovation in Soweto is more a practice to signify the residences presence as habitants of the city and their different class positioning in relation to their broader community. He accurately concludes that the one-dimensional emphasis of the conspicuous element of consumption among the black middle class is a result of racialisation of the meaning of middle-classness (Krige, 2012).

This idea of a racialization of black middle classness is critically examined in how Krige (2012) analyses the works of Brandel-Syrier, who writes about social class in Reeftown. Krige (2012:32) illustrates that Brandel-Syrier narrowly argues that 'the term African

---

[1]One had completed a PhD in the social sciences and another was writing a PhD at the time of the interviews.

middle class can have no meaning in terms of association and social interaction with the European middle class'. Krige takes Brandel-Syrier to task by critically illustrating that the flaws in her argument are homogenising and fixing the experiences or being of the people of Reeftown to their rural past and not to their varied and complex experiences of residing in Reeftown. Krige's critique here is seminal as it is in line with a broader challenge of narrow scholarship that tends to fix the identity and experience of blacks. A similar critique was levelled at Bozzoli's conceptualisation of African societies as being inherently patriarchal compared with white societies (Tshoaedi, 2008). Brandel Syrier's refusal to see the Reef-town middle class as heterogeneous is problematic. Alternatively, in an examination of the American black middle class, Lacy (2007) maintains that to have an unbiased understanding of this group there is a need to make a distinction among the black middle class by looking at their income, wealth, housing, level of education and lifestyle. This will reduce unfounded generalisations and the potential perpetuation of historically racial stereotypes (Lacy, 2007:3). This also allows for an understanding of the complex ways in which the middle class manage their lives when they live among different classes of blacks and in middle class spaces that are racialised.

3. Methodology

This article is a result of a collaborative project on the black middle class.[2] In an attempt to make sense of the concept of the black middle class, this particular part of the study employed the qualitative approach. The article moves from the understanding that the black middle class had been in existence in SA before 1994 (Southall, 2004; Crankshaw, 2005; Mabandla, 2013), although in varied and shifting forms. This consideration is further attested to by the data gathered from the two life histories of women interviewed for this study.

To comprehend the unfolding histories (Hubbard, 2000) of the women, I employed in-depth life-history interviews. The life-history approach was useful in eliciting the patterns of the participants' social relations and processes that shaped them (Bertaux & Kohli, 1984:215). The data can thus be looked at from two perspectives; the lived life that presents the time line that the women shared as they narrated their factual life histories; these factual data are accompanied by the subjective accounts of their lives. These accounts are understood to be located in time and space, thus weaving the storytellers' experiences to the broader socio-economic and political context (Hubbard, 2000).

I was particularly interested in experiences of women who grew up in a middle-class family and were currently positioned in the same class. As a result, the selection of these two participants was purposive and convenient. This choice of sampling was influenced by the fact that there are multiple studies on experiences of class mobility but a dearth of research on a historical analysis of experiences of those born from the black middle class and who continued to be positioned in a similar class. The intention of the study was to get a sense of life experiences from women's perspectives; thus, the findings are not representative of any larger population group but are a personal representation from the women's perspective and my synthesis to the larger social, historical and economic processes that impact the black middle class.

---

[2]The broader project was entitled 'Towards a More Inclusive, Cohesive and Dynamic Society: Understanding the Significance of the Emerging Black Middle Class'.

This is made possible because 'life histories may focus on individual experiences, but that focus does not preclude an examination of social structure' (Hubbard, 2000:4). The synthesis with previous research in the analysis and discussion section of this article attests to this broader examination.

The point of departure for the interviews were to get a sense of how these women got to be middle class by looking back and through sharing their contemporary experiences. The aim was to get accurate descriptions of the women's life trajectories so as to uncover the patterns of social relations, meaning-making and the varied processes that influenced them. Such an understanding provides a deeper explanation of the complexities involved in meaning-making around their being black and middle class.

The interviews were conducted between Johannesburg and Pretoria respectively depending on the availability of the two participants. They were conducted between February and April 2012. Three interviews were conducted with each participant and each interview lasted for one hour to one hour and 30 minutes. All interviews were tape recorded and permission[3] to do so was requested and given prior to the interviews. The life-history interviews provided Aganang and Mosa[4] with an opportunity to share their stories. Regarding who are these women, Aganang answered this question thus:

> I am the fourth child, second daughter in a family of six children; I was born in a small village in the North West Province outside Rustenburg. When I say small village I really mean a small village, because everybody knew everybody else. (Interview, March 2012)

In communities where people intimately know each other, the need to belong is signified; as later narrated by Aganang, this idea of belonging becomes central in how she experiences black middle-classness as shifting and marred with constant complexities. She also reiterated that:

> Although my father was a teacher he also did some farming a little bit of cultivation and little bit of stock farming, so much so that when we went to school he would sell some of his life stock to pay our school fees. (Interview, March 2012)

This information is important in how Aganang thinks about what it was to be middle class and black and her later critique of the Black Diamonds label, discussed later in the article. Mosa answered the same question thus:

Mosa: I would define myself as born and breed in Johannesburg in the township all my life, that's what I would say am a city kid ... I was born at Bridge man but now it's a private hospital and it is in Mayfair, via Brixton – Garden City.

Grace: So it used to be a public hospital? And when was that?

Mosa: 1958! So am not sure if it is because my parents were both black say middle class.? They were public servants my father was trained as a teacher but never worked as a teacher. He worked for the city council of Johannesburg, he was a senior municipal officer and my mother was a nurse at a clinic in Soweto, they both worked in the township, my mother walked to work in Mofolo South clinic my

---

[3]The participants signed a consent form indicating their formal agreement to participate in the study.
[4]To protect the anonymity of the participants pseudonyms are used.

father could walk, but he used to drive to work, he had a car. (Interview, February 2012)

In Mosa's response to who she is there is already a tension seen in her hesitance to define her parents as middle class: 'my parents were both say middle class or what?'. This not fully identifying as middle class is a constant theme throughout the life histories of both women. The rest of the interviews focused on the following themes:

- The beginning (where they grew up, their parents, siblings).
- Their neighbourhoods.
- The socio-economic position of their family in relation to the broader community.
- School years (primary, secondary and university).
- Work experiences.

On further exploration of these general themes opportunities emerged for probing that provided insight into the particular experiences of class at particular life stages. The interviews were illuminating; they revealed an especially interesting history of being black and middle class in apartheid SA. The three themes discussed in this article emerged through the narrating of the women's life histories and were subsequently foregrounded as significant during analysis.

4. Discussion of the two life histories

Most research undertaken on the black middle class is either quantitative or general in that it does not use the life-history approach to understand the phenomenon from the point of view of women. The only work that interrogates women's experiences and negotiations of class positions is the 2009 documentary emanating from the Classifying Soweto Study by Phadi (2010) and her follow-up 2011 article. This limitation as I indicate earlier further influenced my decision to examine women's experiences of being black and middle class.

Both Mosa and Aganang came from middle-class families. Although their self-identification as middle class was critically defined for Aganang because of its impermanence, it was a defining factor in how she experienced her early life. The same is true for Mosa; she suggested that even though being middle class did not openly define how she interacted with her community, the social positioning of her parents facilitated how her life experiences penned out.

4.1 Am I middle class? A critical engagement with the conception of one as middle class

Research on class usually takes it for granted that academic conceptions of the term are similar to those held by the broader society. Again, the characteristics used to denote membership into a particular class are also usually assumed to be uncomplicated, and unquestioningly accepted by those who are theorised about. Recent studies into the middle class, however, have began to illustrate the intricacies involved in self-identifying as middle class, the language used to refer to class and other factors that impact on belonging to and being able to identify with that class (Phadi, 2010; Phadi & Ceruti, 2011 Krige, 2011b). These complexities were clearly visible in how Aganang responded to my request for her to participate in the study. Her response revealed a deep questioning of the usually blind characterisation of particular people as belonging to this class because they share similar features with other members of the class.

When she wrote back questioning whether she was middle class and how I came to that conclusion, I initially thought she was not interested in participating; however, on close attention it became clear that this questioning was embedded in her history and the history of the social, economic and political context of her upbringing and again on her level of education – she had a PhD. She therefore unsurprisingly responded with the following question to my request: 'Am I middle class? What constitutes middle class?'. Her questioning was based on how contemporary discourses on the black middle class tend to emphasise black middle-classness as a post-1994 phenomenon, as linked specifically to conspicuous consumption and thus positioning the black middle-class experience as different to other racial group class experiences. Aganang's critique of narrow conceptions of black middle-classness is a constant thread in her life history. Her narrative rejects an idea of heterogeneous experiences of class by blacks and re-asserts the notion of the existence of a black middle class in pre-1994 SA. These assertions are visible in how her narrative emphasises the shifting and constant negotiation of her class position. In her questioning of her social class position she also said:

> Do we (blacks) belong to a different middle class from other racial groups? If the class division is highly determined by finances, where does social way of life fit into all this? What drives the classification of people as middle class? Is it resources, way of life, occupation? What about all the other kinds of wealth that is not classifiable in the western sense. Simply put where does the western and non-Western approach meet here? (Email communication, February)

Although this quote might initially suggests a view of middle-class experience that is racially homogeneous, at the end Aganang's narration of her life history is told from a restorative stance. Aganang's standpoint here intends to illustrate that apartheid engineering and its thinking of the black experience as homogeneous was not true of her experiences then and now. Her narration was thus both reflective and constitutive as she tries to foreground differential racial and cultural experiences of class from the factual position and from a reflective perspective. Her narration is in line with what Crankshaw (2005) and Krige (2011a) refer to when talking to undifferentiated black experience during apartheid and the conceptualisation of the black middle class as engaged in conspicuous consumption. For example, according to Henderson's (1999) research based in America, racial homogenising is as a result of how, because of the history of racial oppression, low-status and high-status blacks are usually lumped together as an undifferentiated mass. Although this essentialising of the black experience was especially true during apartheid, in contemporary SA race is still signified given how different racial groups experience access as determined by how they were positioned during apartheid (Southall, 2004:522). So the social way of life according to Aganang is still determined by racial disadvantage for the black middle class.

Aganang's response above also raises questions about the assumptions we make as academics when we classify those we study. A similar discussion in Phadi & Ceruti (2011) is used to illustrate the significance of theorising from below and the importance of the historical development of concepts signifying their specific contexts. To illustrate the importance of theorising from below, Phadi & Ceruti (2011: 84) contend that the segregation between academic-conceptual meanings of class from its popular usage-based meanings is not viable in the case of class, because social experiences and the self-conscious articulation of this experience form an important and indispensable aspect of the theoretical concept of class. Thus the conceptualisation of class 'demands the inclusion of popular conceptions'.

At the heart of Aganang's contention to her identification as middle class was her outrage at the public discourse that links the black middle class to conspicuous consumption and the limitation of the Black Diamonds label. Black Diamonds is a concept emerging from market research with a focus on 'individual consumption patterns as an indicator of class (or consumer market segment) rather than on household income or ownership patterns' (Krige, 2011a:297). This narrow conception and its increased use in public discourse on the black middle class is why, to a large extent, Aganang is critical of the black middle-class label in contemporary SA. The label was equally criticised in more progressive analyses of class (Krige, 2011a).

While arguments of conspicuous consumption and the black middle class were present as early as the 1950s in the United States (Frazier, 1957), they are not representative and thus cannot be assumed to represent the middle-class experience of all blacks. Notions that blindly assume that the black middle class engages blindly in conspicuous consumption are a result of stereotyping analysis that fails to critically engage the deep impacts of the homogenising policies of high apartheid, where the class characteristics of blacks were ignored (Crankshaw, 2005), and theorising without an investigation into social meanings of particular practices (Krige, 2011a). Thus Krige (2011a:277) calls for 'more contextual and multi-level interpretations of economic and financial practices and processes such as consumption'.

The Black Diamonds conception was equally questioned in 2005 when it was first put into circulation. This label was not acceptable for Aganang as it seemed too homogeneous and suggested that the emergence of a black middle class was a wholly post-apartheid phenomenon; this is similarly argued by Mabandla (2013). Aganang's reference to the Semes and Motsepes and the question of whether, for our understanding of social position, it counts which social position your parents occupy is important in indicating the historically racialised way class is defined in SA. Again this historicisation is important in how she tells the story of who she is, and how she identified as middle class at times and not at other times. Aganang's response also illustrates her sharp reading of the flaws in contemporary public discourses on the black middle class and its unspoken assumptions that the history of the black middle class is non-existent, insignificant and not to be engaged with. Her contention demands that we broaden our view of black society and thus the different ways of being that racialised access meant for the black middle class, especially during apartheid and still in contemporary SA.

Given the racialised nature of apartheid SA, class was experienced as an uncomfortable and shifting identifier because of its impermanence and convoluted nature in SA, thus leading to a cautious identification with it as a categorisation. The flux of the concept came up in two particularly revealing ways in the two life histories; this impermanence and complexity of class position like in Phadi & Ceruti (2011) study of class in Soweto was identified in comparison with others and with their past. In his theorising on boundaries and class among the black middle class in America, Lacy

(2007) found something similar to what Phadi refers to. He found that middle-class blacks had to not only negotiate their racial identity but had to manage how they interacted with members of lower classes in their own community; these interactions 'shape middle class blacks conceptions of who they are' (Lacy 2007:9). The following subsection captures the two discussions that illustrate how these two women's conception of who they are was influenced by the distinct spaces they occupied and principles they came into contact with in their everyday lives.

4.2 Managing difference and negotiating inclusion

It was confusing for Aganang when she was young to comprehend why her parents insisted that she was not any different from the other children, when she could clearly see that they were different. At the time of the confusion she did not link the differences to class, but it became apparent as she grew older that her father's education and wealth had a lot to do with the difference she felt when growing up. Aganang said:

> the village where I grew up there was this thing were children grew up with their grandparents or sometimes they grew up without parents because their parents were in Johannesburg working but with us we had both parents, that was a source of difference even though my parents especially my father used to hammer the fact that we were not different from the other children but it was funny because whilst he was saying we are no different you see difference every day, you go home there is your mom you go home there is your dad the other kids don't have mom and dad, so how can we not be different. (Interview, March 2012)

The difference she is referring to had to do with the effects the migrant labour system had on families, where children were primarily raised by grandparents. Middle-classness in such communities allowed for the black nuclear family to remain intact, as indicated in the following quote. So for Aganang this meant she was not like everyone else around her and thus she had to constantly negotiate how to be in contradictory ways, as she explained:

> You go home there is a car the other parents don't have cars and father says you are not different you go home come month end your father and mother possibly you too get into this car you go to town to buy groceries you do all these things that the other kids don't do but your father says you are not different, I think what he was instilling in us was humility – be humble don't think you are better off materially, don't think you are better off in any other way. (Interview, March 2012)

This lesson at being humble was an interesting way of negotiating their middle-class position in a community that was not middle class. This parental negotiation of class difference by instilling a certain kind of behaviour was similarly experienced by Mosa. Although like Aganang the difference was present, she did not realise how deep it went in terms of the everyday experiences of those around her. She shared a moment when this difference was made clearer. Mosa said:

> At standard 2 the teacher asked us to talk about something that we did at home and I said in my bedroom I was saying something that I did in my bedroom and the whole class was wow you have your own bedroom it

was strange that nobody had their own bedroom and I was like I thought this was what happens in other families. (Interview, February 2012)

With Mosa's experience, even though the difference was somewhat downplayed even when it was clearly visible in everyday experiences and observation of others around them, it influenced who they became. Again this difference was observed in the food they had access to as compared with the other children. According to Phadi & Ceruti (2011) community-wide comparisons are significant in how class is defined and how one positions themselves in relation to others. This is apparent in how Aganang understood the social position of her family when she was young. Although the food they had access to was supposedly more nutritious, it was different, thus leading to further negotiation of who she was in relation to others. Aganang shares an experience where she and her siblings questioned why their mother did not work in the kitchens so that they could also eat dikokola (dried bread), saying the following to illustrate this point:

> ... we used to have this dikokola, dried bread. Parents who worked as domestic workers took bread that was left over from the table, and would put butter and jam; red jam dried it up (the bread) until the time that they wanted to send it home. There was these kombis or trucks that would take boxes, parcels back home so the mother or the father would send that big box which would have clothing and food. So we had this dried bread which was called dikokola and remember in our village we didn't have a shop, shops were a few kilometres away so we did not have the luxury of having fresh bread so what I remember was that because my mother was not working as a domestic worker my father was not working as a domestic worker we were envious we wanted dried bread we wanted dikokola. (Interview, March 2012)

Comparison and wanting to belong to those around you was experienced deeply by Aganang whose experience of wanting dikokola, which was basically leftover food from the madam's table instead of the fresh bread and dumplings her mother made, was an indication of how being differently positioned as middle class in a non-middle-class environment was somewhat uncomfortably experienced. This explains how the struggle against apartheid was successful in rendering class and gendered struggles insignificant (Ramphele, 2000).

This experience of being different taught Mosa and Aganang not to openly identify with the class position of their family. Given that blacks were restricted to townships and homelands whether they could afford to live elsewhere, they were forced to reside with their own race even though they might have been different with regards to social class. 'Under apartheid, the class divisions within urban African society were "compressed" by a range of policies that ignored occupational class divisions among Africans' (Crankshaw, 2005:354). For example, the 'Apartheid Group Areas Act forced the black middle class to live alongside workers' (Phadi & Ceruti, 2011:93). This is another reason why class was underplayed in black communities during those years and an inclusion that emphasised racial belonging was emphasised instead. This meant their experiences were unlike those of the black middle class in America, who Lacy (2007) suggests engaged in exclusionary boundary work – to illustrate they were different to poor blacks. In the case of Mosa and Aganang they were barred by racial laws that controlled their movement and integration to any other group, and thus they engaged in a more integrationist's negotiation of their class position with that of the various positions of members of their communities. This underplaying of differences of class and gender were undertaken to focus energies on racial oppression (Ramphele, 2000; Krige, 2012). The experiences presented by Mosa and Aganang therefore suggest that these supposedly homogeneous communities experienced class position in complex ways. Mayers (1977, quoted in Phadi & Ceruti, 2011:86) 'illustrates that class stratification manifests in an oppressed environment'. This stratification remains true whether it is acknowledged as present or not.

4.3 Managing difference, negotiating racial exclusion, and being the other middle class

The positions of these two women also illustrate the broader racialisation of relations and social position of the time; they could not openly be middle class in their communities of origin. This was also true when they were in apartheid-engineered white spaces. Even though with regards to class positioning they were similar to most whites, they still could not self-identify with the middle-class position as occupied and lived by whites in similar ways. The privileging of white middle classes during apartheid meant for both Mosa and Aganang that in comparison the privilege and access they experienced in relation to their communities of origin was disadvantaged in comparison with the experiences of the white middle class, thus they were more similar to the black working classes than they could be to the white middle class. This meant they could not even begin to think of what Lacy (2007) refers to as inclusionary boundary work, which means that in America the 'middle class blacks engage in inclusionary-work to establish social unity – to show that middle class blacks are much like the white middle class' (Lacy 2007:76).

This difference is illustrated when Aganang and Mosa went to study at Wits. For Mosa this happened at the height of the students' revolts of the 1970s, which denotes an era where racial difference was signified as compared with deracialised class unity. During high apartheid, this was facilitated by discriminatory state policies (Crankshaw, 2005). This further explains why at this time in their lives they identified more with those back home than their white middle-class counterparts. The difference they experienced when they came to Wits attested yet again to the impermanence and complex class position they were meant to occupy. For Aganang, who had undertaken her undergraduate degree at the then University of the North,[5] coming up to Wits for a postgraduate degree opened her eyes to yet another shift in her ability to identify herself as middle class; she was not the same middle class as the white middle class. Her experience of coming to Wits started with an undervaluing of her degree; she was rejected entry into an honours programme because her third-year degree from Turfloop was viewed as equivalent to a second-year course at Wits. This different valuing was central to apartheid engineering. White and black education was driven by distinct policies, as was access and mobility. Aganang also shares how this was made possible through apartheid state policies:

> The Bantu Education Act was passed in 1953, now the extension of Bantu education Act meant that the Bantu Education Act is extended to

---

[5]The University of the North was popularly known as Turfloop, which was the name of the farm where the university was established. It was renamed Limpopo when it merged with Medunsa in 2005.

> universities where I think people like Dr Motlana and his contemporaries
> studied at Wits but then that privilege or right was taken and therefore we
> couldn't just come. We needed a ministerial permit. (Interview, April 2012)

At the centre of apartheid engineering was a move to strip blacks of ideas of
heterogeneity. Despite the fact that ethnic difference was used to denote difference
among blacks, important analytical differences among ethnic groups like class and
educational background were concealed by how blacks were defined as an
undifferentiated group (Crankshaw, 2005; Grosfoguel, 2004). Again the project to
distinguish blacks through different ethnic groups was mainly to further apartheid,
'determined to divide the non-white majority and secure white supremacy' (Henrard,
2002:49). This attempt at heterogenising blacks was nothing but a continuation of
racialising blacks. Therefore the government, as Aganang explains, did away with
state acknowledgement of African class difference through 'a range of policies that
ignored occupational class divisions among Africans' (Crankshaw, 2005:354). The
absurdity of apartheid policies were also experienced by Aganang in how blacks from
outside SA were provided more privileges as compared with her who was a SA. The
following quote from Aganang indicates how this was experienced:

> now the fascinating thing about the ministerial permit was I am in
> Rustenburg but I need to go to the South African embassy in Mafikeng to
> go and get a permit to allow me as a person from an 'independent'
> country to (laughs) to study, well it's funny now, actually there was a
> policy confusion because if you were from an 'independent' country you
> were supposed to have honorary access. For example the three women
> who were lecturing at the University of Bophuthatswana came here on
> that ticket, so they were guests of the South African Government. They
> lived in one of the residences, they were allowed privileges that your
> regular black Soweto person wouldn't be allowed because look this is a
> white country but I was treated like your regular black because I needed a
> ministerial consent and a ministerial consent would have to state that the
> course that I wanted to do is not on offer in our university. (Interview,
> April 2012)

This quote illustrates the limits of class position when you are not permitted to be in
similar ways as others occupying the same social position; in Aganang's case this was
in reference to the white middle class and middle-class blacks from outside SA who
were given a more privileged position from the one given to her. Phadi & Ceruti
(2011:102) indicate that middle-class identity reflects 'the material reach that social
location confers, but also the width of the social view that different social locations
permit'. On another note, Mosa's experiences illuminated the disparities between the
races much earlier than it did for Aganang. She did her first degree at Wits, and
although her encounter with Wits earlier illuminated her differences to the white
middle class, it meant she was somewhat better able to deal with these differences
when she became a student of the University. Mosa said:

> I applied at Wits, I was send the forms and requested a ministerial consent,
> permission from the minister of Education and training or City Bantu
> education – you couldn't come to Wits because of group areas Act and
> because it's a white university you had to get the permission from the
> minister so I put in this application and I sent it to Pretoria and in the

front they asked why I wanted to come to this university because this degree is not offered in any black universities that was my reason and the minister would consider it. You were never certain that you will get it and then I mean Wits was 99% white there was no black person I think. In my application one of the things they wanted was a testimonial – I needed to get it from my then father in the Anglican Church and he gave me hassles. He said, 'why do you want to go to Wits that is a white university they will never accept you?' I struggled I would go to his office everyday he wouldn't do my letter, he said I don't remember how he said it but he implied that I was not that class that goes to Wits, indicating that people who are from the rich, the well-known people can get there, that is, the Motlana's, we were in the same church with Dr Motlana's family ... (Interview, February 2012)

Mosa's encounter with Mathabatha, one of the individuals active in sensitising black youth during the 1970s in Soweto, meant that her application to Wits was a radical questioning of exclusion, but not necessarily that of inclusionary work as referred to by Lacy. The refusal and questioning by her minister, although a reassertion of class difference within her community, was also an assertion of the precarious nature of her standing in the middle-class position – she and her family were not similar to other 'top' families in her community, thus she could not secure a testimonial from her pastor. The work of challenging exclusion engaged in by the youth of 1976 was based on the idea of a unification of blacks as espoused in the black power movement, and Mosa's participation as a young person at the time influenced her ideas of who she was as a black person rather than as a middle-class black person.

For both Mosa and Aganang their various experiences linked to public encounters with white spaces brought about a re-evaluation of membership to the middle-class position, and somewhat suggested that the idea of difference experienced earlier was not necessarily true or necessary, thus foregrounding their blackness more so than their middle-classness.

Conclusion

Through a detailed discussion of two life histories this article has illustrated that the social position of class, and middle-classness in particular, for the two black women was not experienced homogeneously throughout their lives. The apartheid racial politics and ideas about who you should be among your community and in white apartheid-engineered spaces impacted how they experienced and identified with the label. These findings are significant for our general thinking of class position and the experiences of the black middle class during apartheid and in post-apartheid SA.

In conclusion, this article maintains that being middle class and black during apartheid was filled with complexities. Therefore we need to take into consideration that being middle class and black is heterogeneously experienced and thus should be understood as such.

References

Bertaux, D & Kohli, M, 1984. The life story approach: A continental view. Annual Review of Sociology, Annual Reviews 10, 215–37.

Bhorat, H, Van Der Westhuizen, C & Jacobs, T, 2009. Income and non-income inequality in post-apartheid South Africa: What are the drivers and possible policy interventions? DPRU Working Article, No. 09 (138), Development Policy Research Unit, University of Cape Town, Cape Town.

Crankshaw, O, 2005, Class, race and residence in Black Johannesburg, 1923–1970. Journal of Historical Sociology 18(4), 353–93.

Frazier, EF, 1957. Black Bourgeoisie. The Free Press, Glencoe.

Grosfoguel, R, 2004. Race and ethnicity or racialized ethnicities? Identities within global coloniality. Ethnicities 4(3), 315–36.

Gumede, V, 2011. Social and economic inclusion in post-apartheid South Africa: Poverty and inequality. In Transformation Audit: From Inequality to Inclusive Growth, Institute of Justice and Reconciliation, Cape Town, pp. 88–93.

Henderson, G, 1999. Our Souls to Keep: Black/white Relations in America. Intercultural Press, Yarmouth, ME.

Henrard, K, 2002. Minority Protection in Post-apartheid South Africa: Human Rights, Minority Rights, and Self-Determination. Praeger, Westport, CT.

Hubbard, G, 2000. The usefulness of in-depth life history interviews for exploring the role of social structure and human agency in youth transition. Sociological Research Online 4(4). http://www.socresonline.org.uk/4/4/hubbard.html Accessed 23 March 2012.

Keswell, M, 2010. Education and racial inequality in post-apartheid South Africa. In Attewell, P & Newman, KS (Eds.), Growing gaps: Educational inequality around the World. Oxford University Press, Oxford.

Krige, D, 2011a. Power, identity and agency at work in the popular economies of Soweto and Black Johannesburg. PhD thesis, Wits University.

Krige, D, 2011b. We are running for a living: Work, leisure and speculative accumulation in an underground numbers lottery in Johannesburg. African Studies 70(1), 3–24.

Krige, D, 2012. Histories and changing dynamics of housing, social class and social mobility in black Johannesburg. Interdisciplinary Journal for the Study of the Arts and Humanities in South Africa 19(1), 19–45.

Labov, W, 1997. Some further steps in narrative analysis: Narrative theory. Journal of Narrative and Life History 7, 1–4.

Lacy, KR, 2007. Race, Class, and Status in the New Black Middle Class: Blue-Chip Black. University of California Press, Berkeley and Los Angeles.

Leibbrandt, M, Woolard, I, Finn, A & Argent J, 2010. Trends in South African income distribution and poverty since the fall of apartheid. OECD Social, Employment and Migration Working Article No. 101, OECD Publishing. ww.doi:10.1787/5kmms0t7p1 ms-en Accessed 23 March 2012.

Leibbrandt, M, Wegner, E, & Arden, A, 2011. The policies for reducing income inequality and poverty in South Africa. SALDRU Working Paper Number 64, University of Cape Town.

Mabandla, N, 2013. Lahla Ngubo: The Continuities and Discontinuities of a South African Black Middle Class. African Studies Center, Lieden.

Mason, R, 2007. Conspicuous consumption: A literature review. Emerald Backfiles, European Journal of Marketing 18(3), 26–39.

Phadi, M. (2010). Phakhati: Soweto's Middling Class. Documentary made by Eyelight Productions and the Centre for Sociological Research, University of Johannesburg.

Phadi, M & Ceruti, C, 2011. Multiple meanings of the middle class in Soweto, South Africa. African Sociological Review 15(1), 89–107.

Phadi, M & Manda, O, 2010. The language of class: Southern Sotho and Zulu meanings of middle class in Soweto. South African Review of Sociology (South African Sociological Association) 41(3), 81–98.

Ramphele, M, 2000. Teach me How to be a Man: An Exploration of the Definition of Masculinity. Library of Congress Cataloging-in-Publication Data, The regents of University of California.

Seekings, J & Nattrass, N, 2002. Class, distribution and reconstruction in post-apartheid South Africa: Transformation. Critical Perspective on Southern Africa 50(1), 1–30.

Southall, R, 2004. Political change and the black middle class in democratic South Africa. CJAS/RCEA 38, 3.

Tshoaedi, CM, 2008. Roots of women's union activism: South Africa 1973–2003. Doctoral Thesis, Leiden University.

Visagie, J, 2011. The development of the middle class in post-apartheid South Africa. Presented the Micro-econometric Analysis of South African Data Conference 2011: Salt Rock Hotel. http://www.aceconferences.co.za/MASA%20FULL%20PAPERS/Visagie,%20J.pdf Accessed 23 March 2012.

'Growing up' and 'moving up': Metaphors that legitimise upward social mobility in Soweto

Detlev Krige

A growing body of research on the middle classes in South Africa is concerned with patters of consumption while fewer scholars engage with the inequality that accompanies greater social differentiation. Little existing research addresses the ways in which new members of the black middle class legitimise newfound wealth and social mobility. Taking inspiration from anthropologists who have documented societal responses to changing configurations of wealth and inequality elsewhere in Africa, I employ the life-history method to ask how one Sowetan man, who self-identifies as being black and middle class, frames his own social mobility. I find that spatial metaphors play an important role in legitimising social mobility. I also find that one response to accusations of materialism and conspicuous consumption levelled against the urban black middle class is a renewal of certain cultural practices in which private wealth can legitimately be converted into social wealth.

1. Introduction

A growing body of research on the middle classes in South Africa is reigniting an older debate in the social sciences between competing theoretical conceptions of class, and the methods we use to research class in the contemporary world (Seekings & Nattrass, 2000; Southall, 2004, 2014; Crankshaw, 2005; Schlemmer, 2005; Alexander et al., 2013; Melber, 2013). Seekings (2009) and Southall (2004) have discussed how the two dominant and contrasting approaches to class – neo-Marxist and neo-Weberian – have shaped social science discourse in South Africa (see also Burger et al., 2014). Seekings' (2009) explanation for the dominance of neo-Marxist approaches in the social science literature on South Africa since the 1970s is especially important, as is his call for scholars to familiarise themselves with an earlier 'Weberian moment in South African social science' and to build on an existing 'history of non-Marxist analysis of stratification in South Africa' (2009:881). This earlier Weberian moment was characterised by a number of important sociological and anthropological texts published in the 1960s and early 1970s on urban life under apartheid, specifically Kuper (1965), Brandel-Syrier (1971), and Wilson & Mafeje (1963).

The emerging body of research on changing configurations of class in South Africa draws on multiple traditions. In their edited book on class in Soweto, Alexander et al. (2013) utilise several approaches, with some authors taking an explicit Marxist approach to theorise not only the middle class but also the working class and proletariat (Ceruti, 2013), while others temporarily leave aside Marxist approaches when investigating the linguistic terms Sowetans use when talking about class (Phadi & Manda, 2013) and Sowetans' perceptions of class (Wale, 2013). In his study of the perceptions of social mobility and economic inequality among black individuals in Cape Town, Newcastle and Mount Frere, Telzak (2014) finds varying perceptions about the relationship between race and mobility, what people regard as equitable wealth distribution, and the relationship between education and mobility. He finds that 'perceptions of economic inequality and social mobility are shaped by local economic conditions and by an individual's economic position within his or her community' (Telzak 2014:28). This finding corresponds with arguments presented in the film Phakhati that urban black South Africans' perceptions of class is shaped by their own positioning in a social system, and that they tend to 'class' themselves in relation to their surrounding communities rather than some abstract (national) class model (Phadi, 2013). Their reference groups, then, are decidedly local and social rather than national and abstract.

None of this research, however, addresses the ways in which new members of the black middle class legitimise newfound wealth and social mobility. The only recent study that explores some of the ideological dimensions of the changing class structure in South Africa is the report on 'Middle Classing in Roodepoort' by Ivor Chipkin (2012). Chipkin notices among the black middle class living in urban Johannesburg a process of individualisation, the replacement of the primacy of kin-based relationship with more voluntary forms of association, coupled with the strengthening of kin-based relationships. He notes that this process of individualisation is accompanied frequently by 'an ideology of self-realisation' (Chipkin, 2012:63). The aim of this article is to contribute to this gap in the literature.

Taking inspiration from anthropologists who have documented societal responses to changing configurations of wealth and inequality elsewhere in Africa (Niehaus et al., 2001; Smith, 2001), I employ the life-history method to ask how one Sowetan man, who self-identifies as being black and middle class, frames his own social mobility. As class configurations and wealth distributions in any society changes, so do the meanings associated with class differentiation. Accusations of witchcraft, for example, may function as a levelling mechanism amidst the emergence of wealth inequalities, while it may also be a symbol for new modes of obtaining wealth (Niehaus et al., 2001; Smith, 2001). With that, new ways of legitimising class differentiation and new wealth are likely to emerge. How, then, does my research participant narrate social mobility in his own life story? How does he frame and legitimise such social mobility? How does he respond to growing criticisms of anti-social behaviour among the urban middle classes? How does he reconcile the tensions that arise from short-term, individual material acquisition and its resultant social mobility, with longer term processes of social and societal reproduction, including obligations to kin?

2. Method, mobility, and metaphor

There is little qualitative research on mobility in contemporary South Africa. Apart from earlier sociological work by Brandel-Syrier (1971), Mayer (1977) and Nyquist (1983), and the more recent work by Telzak (2014), Adato et al. (2006), Nieftagodien & Gaule (2012), and Phadi & Manda (2010, 2013), most of the recent literature is quantitative in approach (see Burger et al., 2014). While there is a substantial literature on migration and movement of people across southern Africa, Lee's (2009) recent book African Women and Apartheid: Migration and Settlement in Urban South Africa starts addressing the issue of social mobility and metaphor more directly. Lee traced the ways in which three generations of women adapted to and made Cape Town their home. Innovatively investigating the inter-generational household histories of several families, she traces their migration experiences, employment histories, home renovations and membership of and participation in church groups and voluntary organisations. For the youngest generation of women, 'mobility became something third-generation women actively embraced as a marker of newly won political freedoms and their growing consumerism' (Lee 2009:103; see also Chipkin 2012:67–8). Lee finds that movement has become an important metaphor for mobility that has 'become deeply embedded in women's consciousness and firmly integrated into the text of their lives' (2009:174) – African women have employed these markers of mobility 'as metaphors for a wide variety of transformations experienced in the course of their lifetimes' (2009:180). Like Lee (2009), I too was confronted with the importance of mobility among a young generation of Sowetan fathers during my field research in Johannesburg (Krige, 2010, 2011, 2012, 2014).

For two years I conducted participant observation (see Denzin & Lincoln, 1998) whilst residing in an old yet relatively well-off neighbourhood in 'deep' Soweto. I conducted open-ended interviews, participated in everyday life, became a member of a voluntary association (Krige, 2014) and sought to develop an insider's perspective on social and economic transformations taking place in the largest and most unequal township in South Africa (see Bonner & Segal, 1998; Piazza-Georgi, 2002; Nieftagodien & Gaule, 2012). In my ethnographic engagement with Sowetans, movement and mobility was one of the most important metaphors in discussing social change. While there is evidence in the literature that movement across the cityscape was a marker of social class during apartheid (Brandel-Syrier, 1971), I concur with Lee that mobility has become a metaphor for a range of transformations a younger generation of Africans have experienced. The desire for and energy directed towards mobility by a contemporary generation of Africans is unsurprising, given how apartheid sought to restrict physical movement, and how it limited physical mobilities and cultural crossings (Chipkin, 2012; Krige, 2012).

To explore the theme of mobility more directly, I recently conducted a number of life-story interviews with Sowetan men who self-identify as being black African and middle class, while a colleague followed a similar research method directed at women (Khunou, 2014). We opted for life histories as the research method because we wanted to explore emergent meanings and processes attached to being black and middle class, rather than explore the size of new classes or processes that can be measured with statistical data. Social scientists have recently re-discovered the life history, or biographical method, as one of the tools in their proverbial research toolbox (Denzin & Lincoln, 1998). In a recent article, Fassin et al. (2008) write about violence utilising the biography of a young South African woman, whom they call Magda A. In his response to Fassin et al., Niehaus – who himself has utilised the life-story method in writing about witchcraft, disease and healing in the Mpumalanga Lowveld – writes that 'Life stories enable social scientists to reconcile anthropological and historical concerns and to discern how private domestic relations intersect with the public sphere' (2008:238). Life histories, he surmises, 'are more valuable for that they reveal than for their representativity' (Niehaus, 2008:238). The material collected through interviews I had conducted with one informant – here called Arthur, as he wished to remain anonymous – was chosen because it elicited important facets of how one man narrates his own mobility, and how such narration relates to reference groups that include his close kin as well as the wider society. I have known Arthur for more than a decade, so my interpretation of his narrative is informed by a longer relationship. In this article I am not in the first instance interested in the factual correctness of Arthur's narrative, which is why little effort has been made over and above my familiarity with his circumstances to substantiate events and facts. Again, the strength of the life-history approach does not lie in its representativity, but in what it may elicit.

In this paper I move beyond Lee's use of life histories by linking the use of the concept of mobility in narrated life histories to an important strategy for discursively legitimising upward social mobility, simultaneously silencing its uncomfortable twin – social inequality. I argue that mobility and spatial metaphors legitimise upward class mobility in the life history of Arthur by constantly referencing social relationships (community and kin), and by placing the (linear) life story of the individual into a larger framework of (cyclical) societal reproduction. The process of 'growing up', the most natural phenomenon that transpires between birth and death, is shown to inform and to legitimise another social process – that of 'moving up' or social mobility. I argue also that the metaphors Arthur deploys in telling his own life story potentially render impotent the harshest critique levelled against the black middle class, namely that it is becoming materialistic and that its attachment to consumption is anti-social (see Krige, 2010). In this way, Arthur (consciously or unconsciously) resolves a tension that Parry & Bloch (1989) argue exists in every society: between individual short-term strategies for material acquisition and long-term demands for social reproduction.

In their seminal work on metaphors, Lakoff & Johnsen (2003) argue that the essence of metaphor is understanding and experiencing one thing in terms of another. Structural metaphors – such as 'time is money' – are metaphors where one concept is metaphorically structured in terms of another (Lakoff & Johnsen 2003:7–9). In industrialised societies, they write, time is a valuable commodity because it is a limited resource used to accomplish goals. Because of the way in which the practice and idea of work has developed in recently industrialised societies, work is typically associated with the time it takes, and this time is precisely quantified because people are paid by the hour, week and year. The association of time with money, and the associated practices (air time, wages, hotel room rates, interest on loans, etc.), is relatively new in the history of the human race. Orientational metaphors have to do with spatial orientation: up–down, in–out, front–back, on–off, deep–shallow, central–peripheral (2003:14). Resulting from the fact that we have 'bodies of the sort we have and that they function in our physical environment', orientational metaphors give a concept a 'spatial orientation' (2003:14). For example, 'happy is up'. The spatial orientation of metaphors is not arbitrary: 'They have a basis in our physical and cultural experience. Though the polar oppositions up-down, in-out, etc., are physical in nature, the orientational metaphors based on them can vary from culture to culture' (2003:14).

Lakoff & Johnsen claim that most fundamental concepts are organised in terms of one or more spatialisation metaphors, and that there is an internal systematicity to each spatialisation metaphor (2003:17). In some cases, they argue, 'spatialisation is so essential a part of a concept that it is difficult for us to imagine any alternative metaphor that might structure the concept. In our society 'high status' is such a concept' (2003:18–9). The 'high status' metaphor, they write, has both a social and physical basis in that status is correlated with social power and physical power. Not all societies associated 'high status' with 'up', but typically physical and cultural experience provides the possible bases for spatialisation metaphors. We will see how Arthur deploys spatialisation metaphors in telling his own life story, and in articulating his understanding of social learning and social mobility. In making sense of social mobility and an abstract notion of 'the middle class', Arthur deploys a range of spatialisation metaphors that at once frames the situation and provides a legitimisation for his own position. These spatialised metaphors speak to the transition from an urban–rural nexus to a township–suburb one; becoming conscious of social and cultural distances as he is exposed to a multicultural learning environment; the way his primary social group moves through time and space; how status in society is conceptualised; and movement metaphors revealing occupational and lifestyle changes.

3. The family unit, the rural–urban nexus and education

Arthur was born in 1971 in what is today known as Limpopo and spent the first few years of his life living in a peri-urban area in this northern region of South Africa. By 1975 his mother was working as a teacher in a Sowetan school while Arthur and his siblings remained in the then Northern Transvaal because she did not qualify for a municipal house. Later that year she was given 'the rights and the papers' to a municipal house, which she occupied in December 1975. After successfully negotiating the bureaucratic process of getting permits for the children to stay with her, they joined her in 1976.

Arthur was enrolled in Grade One in 1976; however, between June and August no schooling took place because of the chaos that accompanied the June 16 uprisings (Bonner & Segal, 1998; Chipkin, 2012:64). In September 1976, Arthur and his siblings returned north and it was only during 1979, 'when things started cooling off', that they returned to their mother in Soweto: 'Then after coming back already the standards were a little bit different [than at school in the north]'. His mother held on to her job, the only thing that could secure her right to the city given that she was not born in the city (Posel, 1991). He finished his schooling in Soweto and like most other youths his education was affected by the various states of emergencies that rocked South Africa in the mid-1980s. Little schooling took place during these periods, yet Arthur passed all his subjects, and in 1987 the family started considering moving him to another school, after hearing of good schools opening up their doors for black African students. In his case, Arthur told me, it was possible because of the way in which he excelled in sport, specifically tennis and athletics.

Arthur applied to a private college in Johannesburg, took their entry examinations and was admitted on the condition that he repeats Grade 10 (Form Three). In 1987 he enrolled in the mixed-race college, going on to complete Grade 11 in 1988 and studying for his Matric in 1989. He was awarded a sports bursary at the college, where he completed his high school, and later received another bursary for tertiary studies at a Johannesburg university where he spent three years studying towards a degree in BCom Marketing before the company sponsoring his studies discontinued their bursary out of fear for the political changes that may have resulted from the first democratic elections that was to take place in 1994. He thus experienced some of the contradictions of political freedom – the doors of learning being opened but the funding for such learning drying up. He continued his studies at Vista University in Soweto by enrolling for a BA degree in 'economics, public administration, and political science', completing it in 1997.

The outline of this part of his life that Arthur sketched was filled with comparisons, movement across time and space, and a narrative employing a series of spatial and social metaphors that both underplayed and emphasised his own agency in the processes of social mobility and inter-generational reproduction. As the child of a teacher, education was central to Arthur making sense of his own life trajectory:

> Education played a very, very important role ... but as you grow you find that the one thing that really kept us going – or [kept] the fire burning – was that because we were raised by a teacher and a principal, we were really reminded that if we don't get education our life will never change [...] despite the circumstances of the then time, you find that when you move from one part of the country into the other part of the country, that's exactly when your life changes. I mean growing up in Limpopo [...] you then again grow up in Johannesburg, Soweto, then you go back and forth.

His physical prowess in various sports certainly played a role in consolidating a strong 'spatial orientation' of personal and social mobility, and possibly a masculine one. His narrative is cognisant of the fact that life opportunities increased massively as Africans migrated to urban areas (Bozzoli, 1991; see also Louw et al., 2007) and the way in which his movement between the urban–rural nexus structured his earliest reference groups. It also acknowledges that his life opportunities increased as he attended a mixed-race school:

> It's important. It gives you exposure. Exposure to the language, exposure to culture, exposure to other people's dreams, and exposure to other people of colour who have or possess what you do not have. You are a bursary student. You don't come to school with a car; you come with bus or a train. And you meet other people who they take you to their homes over the weekends. You look at the type of food they eat, what their father does, what their mother does, it's a total different aspect. It's like moving me from the North Pole to the Equator at the same time.

Even as he was able to enjoy the privilege of attending a mixed race school, Arthur was conscious of his own position in relation to other students, and the different class positions they occupied. This class and social distance that was at once very personal yet public is also spatialised and naturalised with reference to a global spatial metaphor (the North Pole and the Equator), while cultural and gendered distances are spoken of as physical and geographic distances. While the distance is great, such movement is indeed possible.

4. The world of work, wages and home ownership

In 1997 Arthur entered the world of work as a graduate trainee at an information technology company in Johannesburg, but had to leave after two years because the company 'went down'. With the retirement savings ('pension monies') he decided to collect on his exit, he opened a hair salon and a shop in Soweto, employing four hairdressers and one receptionist. The enterprise was robbed at gunpoint twice in the space of a year, and because the salon was closed at times for weeks on end before opening again, he started losing clients and eventually he had to close (down) the business. After several years of working at various companies, with periods of unemployment in between, by 2004 he was working for 'one of the biggest then companies in IT [information technology]' – and it is with the accompanying increased income that he could claim to have joined the black middle class:

> That's when I started. I bought my house but I already had cars. I bought land and I built a house with an X [sic] amount of money invested into it. And then having to finish the house with the furniture, then having to also up the type of car that I used to drive.

It was at the beginning of this period in his life that I met Arthur. At the time of our first formal interview a decade later, Arthur had just found employment again after a year of being unemployed. This is why he could express, during the interview, that he is successful in his working life. But a few months prior to our interview he had in private expressed much despair and angst, as the bank was about to repossess the house he had built in an upmarket Sowetan neighbourhood for falling behind on repayments. He had lost his one car and was hurting from a ruinous marriage. But in 2013 'things were looking up'. While he was quick to say 'I was never born middle class', despite his mother being a teacher, he did not hesitate to say that he joined the middle class: '[When] I started owning my own house'. Ownership of a car at a previous point in his life was not as important as this moment: 'My house, that's how you define me because of the bond, the furniture'. The social science literature attests to the importance of private home ownership as a symbol of class in Soweto (Parnell, 1991; Ginsberg 1996; Morris et al., 1999; Crankshaw et al., 2000; Krige, 2012), even if economists disagree about the relationship between home ownership, labour market mobility and social mobility.

Arthur was proud that he had purchased the furniture with cash, not on credit (see Phadi & Ceruti 2013:156). Following this statement he immediately recounted, as if to counter possible accusations of individual acquisitiveness, how he also spent money on purchasing a tombstone for his late mother's grave. In much of Sowetan society, the close kin of someone who has passed on has the responsibility of erecting a tombstone on the deceased's grave after a period of mourning has lapsed. This 'unveiling of a tombstone' is an important ritual in ensuring that the deceased's relations with the living is peaceful, and that the deceased is on the way to becoming an ancestor. Conflicts within families, economic and social problems, and ill health may all be the result of kin not having fulfilled their responsibilities towards a deceased family member (Ashforth, 2000). Similarly, the 'unveiling of a tombstone' may be an expression of gratitude towards an ancestor for bestowing spiritual and material well-being. In this manner, Arthur inserted his short-term acquisition, which brought him into the middle class, into longer term social processes that include his wider kin group. His narrative allowed his personal wealth to be converted into social wealth. A range of similar cultural practices have found a ready audience among some young members of the urban black middle class, ranging from various 'thanks-giving ceremonies' for various markers of social upward mobility (getting a job, buying a car, relocating to the suburbs, etc.) to more established events signalling a change in social status (birthdays, initiation, graduation, marriage, death).

As a self-identified 'traditionalist', Arthur commented:

> The house is the only [material] thing otherwise everything [that] was supposed to be done [was done]. I mean my mother's tombstone; that was

a bit of traditional move that I did, I had to do it because I believe that I couldn't have waited.

After he qualified for the bond, bought the property and moved into the house he had built on it, Arthur consulted with his father about the design and purchase of a tombstone for his late mother's grave: 'I did it at that time because I believed that I wouldn't like the old man to pass on without having done that'. Seen in this light, erecting a tombstone is not merely a short-term decision about purchasing a commodity with a short lifespan; it is also an investment in longer term cycles of social reproduction related to one's kin group that promises social, physical and economic health. Arthur made this point explicit by saying: 'So it represents a great, bigger scheme of things. Ja, it's not only for me, it's for everybody and then now it's there and it will – it's time immemorial. I mean it stays, stays, stays there'.

Anthropologists Parry & Bloch argued that all societies need to make:

> some ideological space within which individual acquisition is a legitimate and even laudable goal; but that such activities are consigned to a separate sphere which is ideologically articulated with, and subordinated to, a sphere of activity concerned with the cycle of long-term reproduction. (1989:26)

They argue that the maintenance of the long-term order (and social reproduction generally) is both 'pragmatically and conceptually dependent on individual short-term acquisitive endeavours' (Parry & Bloch 1989:26). The ancestors of Arthur, including his deceased mother, were dependent on Arthur's material acquisitiveness and upward social mobility for becoming proper ancestors. Simultaneously, investing in a tombstone was a way in which Arthur could articulate his own material acquisitiveness with social processes that transcended his individual life, and included his living and dead kin. Importantly, this process of articulation legitimised the acquisition of new wealth in public too: among residents of Soweto, friends and a more abstract public in which accusations of materialism have become an important trope for talking about the black middle class.

In Arthur's conceptions of being middle class, ownership and access to material objects of consumption – which are also markers of modernity (Appadurai, 1995) – are necessary, but not a sufficient condition for being middle class. That the purchase, use and exhibition of different types of consumables such as food and cars play a role in the construction of social class registers goes without saying (see Meintjes, 2000). As important is the ability to establish new relationships of hierarchy and dependence (see Ferguson, 2013), including the ability to give and not only take wages. Recounting in the present tense his material situation of a few years ago, before he became unemployed, Arthur said:

> Back home I've got a computer, I mean I'm connected, I'm wired. How many people are wired? You know, I've got an office at home. I mean I've got two fridges, once in my time I had five cars. I wouldn't really doubt myself to say that I am a middle-class type of a person. I used to have a domestic worker. I have a gardener who comes every week and does my garden. Hence, I said I've got a boy who used to come and wash my cars. I am paying a salary to somebody or I'm paying a wage. If I was to be sick or dead, those three people wouldn't get a salary from me.

This was the only instance during our interviews that Arthur self-identified so strongly as being middle class – and for him an important part of middle-classness was being a source of income for others through offering wage work, and thus entering into new hierarchical relationships and dependencies with people (see Dilata, 2008).

5. Inter-generational mobility and barriers to an expanding reference group

This brief overview of Arthur's engagement with the world of work should already give us a sense of the way he moved – in the space of a decade – inside and outside and between the various categories that we use to make sense of the 'market for labour'. They constitute moments of highs and lows, achievement and loss. Narratives such as Arthur's had become quite familiar in Soweto. Those who had moved out of the township 'with a better-than-you attitude' at the height of their earnings were often scorned by working-class men and women on their return to their mothers' houses in the township after they had fallen on hard times. What happens to your self-identification as middle class during these momentary or prolonged periods of 'in-between', when you try and hide your unemployment and even poverty (Meintjes, 2000)? Arthur was adamant:

> No. You don't move out of the middle class. As much as you still occupy that space – the house, the car. I moved from a bigger BMW to a [smaller] BMW. You will say that it's still a BMW. Do you know what I'm trying to say? I still cook the same way. Ja, I had the same type of – even though certain things were not there but I did enjoy the comfort zone of that type of a thing, I mean the clothes everything ... I think anybody who has actually advanced from a basic type of a lifestyle into a basic luxurious lifestyle, they will never go back. They will hustle to keep it there because remember you are not an upper class. I mean moving from a middle class to [not] – it's quite a stretch; it's not as easy as we move from basic to middle class, to lower class.

Arthur was implying that the movement from middle class to lower class was not as drastic as the move from upper class to unemployment. This led us to talk about what Arthur called the 'yo-yo movement', the upward and downward movements of individuals and households between social class categories. In his case he was expressing the dynamics of both inter-generational and intra-generational mobility. Whereas inter-generational mobility compares parents' social class levels with that of their children, intra-generational mobility plots social class shifts within a person's lifetime. The resulting movement is called social mobility – which can be either upward or downwards – and occurs whenever people move across social class boundaries. Again, this movement is expressed using spatialised metaphors that also index the social: 'leaving the township', 'moving to the suburbs' and 'returning to your mother's home'.

Several barriers made it difficult for Arthur to conceive of his reference groups as inclusive of members of the white middle class and to conceive of a grouping that could constitute a national middle class with some shared interests. He noted: 'I would mostly compare myself to black people and to the people that I know they've gone to school, they are working, and then they are doing something about themselves'. Racial inequality was not only the result of apartheid history, but was exacerbated by practices within the corporate world that included salary scales, the culture of the workplace, and informal networks of power and influence:

> No, there is no national middle class because an average white graduate and an average black graduate will never earn the same ... Even today [...] In

the corporate industry position advancement, and training, and remuneration of white people, it's been tailored differently to black people.

Company managers were using arguments about property prices, rent and living expenses being cheaper in the townships as a reason for paying black managers lower salaries (echoing Bundy's argument about how the owners of gold mines justified low wages to migrant labours because they were said to be supported by subsistence economies in the homelands; cf. Bundy, 1979). So even if you are employed on the same level ('bracket'), the white worker is likely to earn more. Whereas the white worker stays close to Sandton and commuting to work takes him only 15 minutes, black workers living in the township have to travel 45 kilometres. What economists would call the 'asset deficit' of the black middle class (Burger & Van der Berg, 2004) plays out also in terms of other aspects of their lives. Racial inequality is spatialised in that where you live and how far you have to travel to work opportunities structure your income, expenses and mobility.

His salient views on continued racial inequalities as a barrier to conceive of a national middle class was complemented by the economic impact of cultural practices such as high costs relating to marking changes in social status among black people (specifically marriage) and to important variations in inter-generational wealth transfers between black and white households. It was expected of Arthur as he established himself as middle class to transfer wealth upwards on a generational scale – towards his living parents by repaying study loans ('payback is a bitch'), supporting them financially ('you leave the home spic and span'), and eventually towards his deceased ancestors by successive thanksgivings. Anecdotal evidence suggests that when it comes to white households that have been middle class for one or two generations, the direction of the inter-generational transfers of wealth is reversed: parents typically transfer wealth to a younger generation through ritualised gift-giving: a motor vehicle for a 21st birthday; a bed as wedding gift; or a washing machine upon buying a new home.

6. Concluding remarks

There are many aspects of Arthur's life story I did not include in this discussion, including his marriage and his relationship with his children. As someone who describes himself as a 'traditionalist', Arthur does not belong to a family that have known wealth. While his mother was a teacher, he did not consider himself being born into the middle classes. Within two generations, Arthur's family unit migrated from a semi-rural context to an urban one, retaining kinship connections in both contexts. The rural context remained an important reference point throughout his life. His household transitioned important political and economy eras, from the racial capitalism of apartheid (Terreblanche, 2003) to a liberal constitutional democratic order – and both educational and employment opportunities opened up as a result of this political shift. But he also experienced early on how this shift closed down other opportunities for him. Nonetheless, he remained successful, avoiding in the process the shortened life that befell many of his generation of Sowetan men. His education history is one characterised not only by what he had learned, but how and where – tracing the ways in which race and class have been spatialised. It documents his movement from a space of social homogeneity and greater similarity to social heterogeneity and unsettling differences.

He moved between various kinds of employment, from being self-employed and employing others in a capitalist enterprise, to being a manager in a corporation.

Importantly, he also stood in an employer relationship to a gardener and domestic worker, and his ability to give wages and not only takes wages allowed him to renegotiate social hierarchies not in his place of work but in his residential and social world. It is within these local contexts and reference groups that the metaphors of upward mobility he employs have their strongest resonance. He traversed several spaces within Soweto and Johannesburg, from living with his parents to renting property and becoming a property owner in an upmarket Sowetan neighbourhood. He was self-employed in the township; and became a manager in the suburbs. He was a wage-giver in the township and a wage-taker in the suburbs. The social mobility resulted in a complex process of constantly having to negotiate new and older relationships of dependence and hierarchy across spaces. Like being an employer, owning property was a decisive threshold for him in the process of 'growing up' and 'moving up'; climbing a class ladder that is intimately linked to his social status in Sowetan society

Arthur's life story is characterised not only by inter-generational mobility, but also by intra-generational mobility as he moved from being a student to employment, self-employment, unemployment and then becoming a manager in a corporation. His own trajectory between different kinds of occupations and levels of employment is not unique, and he compares it with the 'yo-yo movement' that describes the movement of Sowetans as they negotiate different class positions over time depending on their income and residence arrangements. This movement he also linked to a more public process that took place during the early 1990s when some Sowetans moved from 'eating pap and meat and cabbage to sitting in Parliament'.

In telling his life narrative, Arthur made an effort to turn every event that marked upward social mobility into an opportunity to connect with the social space of his kin group and the obligations he feels he had towards them. The purchase of a new car, landing a new job, moving into a new home are all events that allow for, if not require, a re-adjustment to the social world given new hierarchies and new positions. Thus 'thanksgiving events', including 'tombstone unveilings', are also events that allow for the working out of new social hierarchies and new behaviours. In this admittedly 'traditionalist' way, reconnecting one's personal mobility and new standing with enduring social groups such as kinship and residential communities also legitimises one's new standing; and personal wealth is transformed into social wealth. This is not to say that Arthur is insincere in performing these ceremonies, that their legitimising function is the reason for conducting such ceremonies in the first place. Rather, it is to recognise the legitimising consequences of such social ceremonies in a society where social relations, in additional to class relations, remain important for working out social hierarchies and for negotiating upward social mobility. Spatial metaphors play a central part in giving expression to this social mobility that intersects with physical mobility at times, while these mobilities are legitimised through metaphors that reference social relations rather than abstract categories.

Acknowledgement

The author would like to acknowledge research funds received from the University of Pretoria's Research Development Grant as well as the University of Stellenbosch Research Division (Subcommittee A). The author would like to thank two anonymous reviewers for their constructive reviews, as well as the following people for providing insightful comments on earlier drafts: Guillame Johnson, Grace Khunou, James Merron and Nozipho Mngomezulu.

References

Adato, M, Carter, MR & May, J, 2006. Exploring poverty traps and social exclusion in South Africa using qualitative and quantitative data. The Journal of Development Studies 42(2), 226–47.

Alexander, P, Ceruti, C, Motseke, K, Phadi, M & Wale, K (Eds), 2013. Class in Soweto. University of KwaZulu-Natal Press, Scottsville.

Appadurai, A, 1995. Modernity at Large. University of Minnesota Press, Minnesota.

Ashforth, A, 2000. Madumo, A Man Bewitched. University of Chicago Press, Chicago.

Bonner, P & Segal, L, 1998. Soweto: A History. Maskew Miller Longman, Cape Town.

Bozzoli, B, 1991. Women of Phokeng. Ravan Press, Johannesburg.

Brandel-Syrier, M, 1971. Reeftown Elite. Routledge & Kegan Paul, London.

Bundy, C, 1979. The Rise and Fall of the South African Peasantry. University of California Press, Berkeley.

Burger, R & van der Berg, R, 2004. Emergent black affluence and social mobility in post-apartheid South Africa. Working Paper 04/87, Development Policy Research Unit, University of Cape Town.

Burger, R, Steenekamp, C, van der Berg, S & Zoch, A, 2014. The middle class in contemporary South Africa: Comparing rival approaches. Development Southern Africa 32(1). DOI: 10.1080/0376835X.2014.975336

Ceruti, C, 2013. A proletarian township: Work, home and class. In Alexander, P, Ceruti, C, Motseke, K, Phadi, M & Wale, K (Eds), Class in Soweto. University of KwaZulu-Natal Press, Scottsville.

Chipkin, I, 2012. Middle Classing in Roodepoort: Capitalism and Social Change in South Africa. Public Affairs Research Institute, Johannesburg.

Crankshaw, O, 2005. Class, race and residence in Black Johannesburg, 1923–1970. Journal of Historical Sociology 18(4), 353–93.

Crankshaw, O, Gilbert, O & Morris, A, 2000. Backyard Soweto. International Journal of Urban and Regional Research 24(4), 841–57.

Denzin, NK & Lincoln, YS (Eds.), 1998. The Landscape of Qualitative Research. Sage, London.

Dilata, X, 2008. Between 'sisters': A study of the employment relationship between African domestic workers and African employers in the townships of Soweto. Unpublished MA thesis, Industrial Sociology. University of the Witwatersrand, Johannesburg.

Fassin, D, Le Marcis, F & Lethata, T, 2008. Life & times of Magda A: Telling a story of violence in South Africa. Current Anthropology 49(2), 225–35.

Ferguson, J, 2013. Declarations of dependence: Labor, personhood, and welfare in Southern Africa. Journal of the Royal Anthropological Institute 19, 223–42.

Ginsberg, R, 1996. Now I stay in a house: Renovating the matchbox in apartheid-era Soweto. African Studies 55(2), 127–39.

Khunou, G, 2014. What middle class? The shifting and dynamic nature of class position. Development Southern Africa 32(1). DOI: 10.1080/0376835X.2014.975889

Krige, D, 2010. Looking at inequality and class through the drinking glass: An ethnography of men and beer-drinking in contemporary Soweto. In Van Wolputte, S & Fumanti, M (Eds.), Beer in Africa. Verlag, Berlin.

Krige, D, 2011. Power, identity and agency at work in the popular economies of Soweto and Black Johannesburg. Unpublished DPhil thesis, Social Anthropology, University of the Witwatersrand, Johannesburg.

Krige, D, 2012. The changing dynamics of social class, Mobility and housing in black Johannesburg. Alternation 19(1), 19–45.

Krige, D, 2014. 'Letting money work for us': Self-organization and financialisation from below in an all-male Savings Club in Soweto. In Hart, K & Sharp, J (Eds.), People, Money and Power in the Economic Crisis: Perspectives from the Global South. Berghahn Books, London.

Kuper, L, 1965. An African Bourgeoisie: Race, Class and Politics in South Africa. Yale University Press, New Haven, CT.

Lakoff, G & Johnsen, M, 2003. Metaphors We Live By. University of Chicago Press, London.

Lee, R, 2009. African Women and Apartheid: Migration and Settlement in Urban South Africa. Tauris Academic Studies, London.

Louw, M, Van Der Berg, S, & Yu, D, 2007. Convergence of a kind: Educational attainment and intergenerational social mobility in South Africa. South African Journal of Economics 75(3), 548–71.

Mayer, P, 1977. Soweto People and their Social Universe. HSRC Press, Pretoria.

Meintjes, H, 2000. Poverty, possessions and 'proper living': Constructing and contesting propriety in Soweto and Lusaka. MA thesis, Social Anthropology, University of Cape Town, Cape Town.

Melber, H, 2013. Africa and the middle class(es). Africa Spectrum 48(3), 111–20.

Morris, A, Bozzoli, B, Cock, J, Crankshaw, O, Gilbert, L, Lehutso-Phooko, L, Posel, D, Tshandu, Z & Van Huysteen, E (Eds), 1999. Change and Continuity: A Survey of Soweto in the Late 1990s. University of the Witwatersrand, Johannesburg.

Nieftagodien, N & Gaule, S, 2012. Orlando West Soweto: An Illustrated History. Wits University Press, Johannesburg.

Niehaus, I, 2008. Comment. Current Anthropology 49(2), 238–9.

Niehaus, I, Mohlala, E & Shokane, K, 2001. Witchcraft, Power, and Politics: Exploring the Occult in the South African Lowveld. Pluto Press, Cape Town.

Nyquist, TE, 1983. African Middle Class Elite. Occasional Paper number 28, Institute of Social and Economic Research, Rhodes University, South Africa.

Parnell, S, 1991. The ideology of African home-ownership: The establishment of Dube, Soweto, 1946–1955. South African Geographical Journal 73(2), 69–76.

Parry, J & Bloch, M, 1989. Introduction: Money and the morality of exchange. In Parry, J & Bloch, M (Eds), Money and the Morality of Exchange. Cambridge University Press, Cambridge.

Phadi, M. 2013. Phakhati. Eyelight Productions, Johannesburg.

Phadi, M & Ceruti, C, 2013. Models, labels and affordability. In Alexander, P, Ceruti, C, Motseke, K, Phadi, M & Wale, K (Eds.), Class in Soweto. University of KwaZulu-Natal Press, Scottsville.

Phadi, M & Manda, O, 2010. 'The language of class': Southern Sotho and Zulu meanings of 'middle class' in Soweto. South African Review of Sociology 41(3), 81–98.

Phadi, M & Manda, O, 2013. The language of class: Confusion, complexity and difficult words. In Alexander, P, Ceruti, C, Motseke, K, Phadi, M & Wale, K (Eds.), Class in Soweto. University of KwaZulu-Natal Press, Scottsville.

Piazza-Georgi, B, 2002. Human and social capital in Soweto in 1999: Report on a field study. Development Southern Africa 19(5), 615–39.

Posel, D, 1991. The Making of Apartheid. Oxford University Press, Oxford.

Schlemmer, L, 2005. Lost in transformation? South Africa's emerging African middle class. Occasional Papers No 5, Centre for Development and Enterprise, Johannesburg.

Seekings, J, 2009. The rise and fall of the Weberian analysis of class in South Africa between 1949 and the early 1970s. Journal of Southern African Studies 35(4), 865–81.

Seekings, J & Nattrass, N, 2000. Class, distribution and redistribution in post-apartheid South Africa. Transformation 50, 1–30.

Smith, DJ, 2001. Ritual killing, 419, and fast wealth: Inequality and the popular imagination in southeastern Nigeria. American Ethnologist 28(4), 803–26.

Southhall, R, 2004. Political change and the black middle class in democratic South Africa. Canadian Journal of African Studies 38(3), 521–42.

Southall, R, 2014. Black and middle class in South Africa: 1910–1994. Paper presented at the Wits Institute for Social and Economic Research, University of Johannesburg, 10 February.

Telzak, SC, 2014. Trouble ahead, trouble behind: Perceptions of social mobility and economic inequality in Mount Frere, Eastern Cape and Newcastle, KwaZulu-Natal. CSSR Working Paper No. 326, of Cape Town.

Terreblanche, S, 2003. A History of Inequality in South Africa, 1652–2002. University of Natal Press, Durban.

Wale, K, 2013. Perceptions of class mobility. In Alexander, P, Ceruti, C, Motseke, K, Phadi, M & Wale, K (Eds.), Class in Soweto. University of KwaZulu-Natal Press, Scottsville.

Wilson, M & Mafeje, A, 1963. Langa: A Study of Social Groups in an African Township. Oxford University Press, Cape Town.

Food, malls and the politics of consumption: South Africa's new middle class

Sophie Chevalier

Consumption has become a central focus in South African politics, one that hinges especially on evaluation of the behaviour of the new black middle class. Based on an ongoing ethnographic study of Durban, mainly among the lower middle or 'professional' class across a range of racial categories, the article addresses three aspects of this question: food provisioning and consumption across and within the various communities; interaction in shared social spaces that were previously segregated, especially shopping malls; and moral discourses in the media concerning this new class. The so-called 'black diamonds' are a South African urban type of the sort labelled by Benjamin as a phantasmagoria. South Africans are willing to experiment beyond the boundaries of their native communities and there is an emergent national middle-class culture, but there are marked regional differences and nothing yet that would amount to 'creolisation'.

1. Introduction

Since 2008 I have carried out intermittent ethnographic research in eThekwini/Durban (hereafter Durban) on the new middle class, across a range of racial categories; and for the last three years one major focus has been on food provisioning and consumption. I am interested, as are most South Africans, in the relationship between race and class, and the possible emergence of a non-racial middle class. The growth of a new middle class of consumers has been and still is taken as a measure of success in transforming the country's society and politics. If a middle class is not made by how it spends money (consumption) but by its sources of income and property, consumption has certainly been used to define individual members of such a class.

When I started out, I decided to focus my attention on the two features commonly taken in public discussions as signs of the emergence of a new non-racial middle class: access to consumption patterns previously reserved for whites; and residential mobility – that is, households that left the townships for the formerly white suburbs. Much of my subsequent research has been concerned with social interaction in shared public spaces that were segregated under apartheid: residential areas and car transport, use of the sea front (beach and promenade) and shopping malls by all classes and communities as free spaces (e.g. Chevalier, 2012). In the process I extended my initial focus to include all racial communities. This article first looks at consumer behaviour, approached more narrowly through the study of food provisioning; second, I address the role of shopping malls in developing new patterns of consumption and social interaction; and, finally, I take up public discourse about the new middle class, as represented by discussion of the so-called 'black diamonds'. For South Africa, I argue, this is a new 'urban type' in Benjamin's (2006) sense of a phantasmagoria.

My general hypothesis is that the new multi-racial middle class recreates the South African nation through their consumption, while at the same time making new links to world society. A vast new arena of consumer choice, which was the preserve of whites not long ago, has been opened up to blacks – albeit often on credit. This offers the latter an opportunity to participate in national 'prosperity', while their fascination with international brands reflects a similar desire to participate in global consumption. My work also aims to add a neglected regional dimension to marketing and media discussions of race and class. Marketing specialists treat South Africa as a single entity stratified before by race and now by social class. Durban offers an interesting commentary on these generalisations because of its specific social history, combining the former British colony of Natal, a strong Indian presence (it is the largest Indian city outside India) and the home of the Zulus, South Africa's largest ethnic group.

My research has been based in part on visiting individual shops and malls, on interviews with shopkeepers and with families at home, mostly women from all communities.[1] Fieldwork was carried out in 2011 (six weeks), 2013 (four weeks) and 2014 (six months). I contacted my middle-class informants through personal networks based on neighbourhood, schools and workplaces. I met some of these women several times; and I exchange a number of items with them when I am in Durban – recipes, tips about cookery books, shops addresses and foodstuffs. I begin here by summarising my interviews with women on shopping, food consumption and cooking. I have also consulted numerous media sources that offer advice to consumers, such as television programmes, cookery books and women's magazines (two of which, *Bona* and *Your Family*, I am studying intensively).

2. Food consumption

Studying food consumption is for me part of a long-term research project, which includes urban food supply chains – markets, supermarkets, and so forth – and is comparative in scope (on France and Britain, see Chevalier, 1997, 2007; on South Africa, see Chevalier & Escusa, 2011). Looking at food allows us to integrate a wide range of questions because, although we need to eat to live, eating is also a principal vehicle for sharing social life and for expressing a range of lifestyles. Moreover, food is a deep symbol of inherited identities. Every group has its 'soul food' and the taste for this changes more slowly than our willingness to experiment with fast foods imported from around the globe. Without doubt, the residents of cities such as Durban today have a vastly expanded repertoire of foodstuffs at their disposal. Yet this development, far from leading to a homogenisation of food regimes, has allowed people everywhere to push out the boundaries of their culinary experience, while still defining their own authenticity in quite narrow ways. People go out to buy food, both raw and cooked; they consume it at home or in cafes and restaurants, with family and friends. By

---

[1]Until now I myself have interviewed members of some 40 households once or more times. One-half of them were black, the others Indian, white and coloured (two were multi-racial couples). With few exceptions, they had children; and one-third of these households included extended kin. They were all between 30 and 50 years old, but they distinguished themselves as generations by whether they came to maturity before or after the early 1990s.

focusing on this movement of people as consumers, linked to what they eat, where and how, we may be able to reveal complex patterns of interaction across public and private spheres, as well as old and new forms of social, ethnic and religious identity. Clearly, food is 'good to think with' when we contemplate the transformations now taking place in South Africa.[2] Middle-class consumption, especially for the lower middle class that I have focused on, can also be precarious, even fragile; and nothing illustrates this more vividly than fluctuations in food-buying habits.

2.1 A tight food budget

My material shows that everyone took great care when budgeting for food and often observed strict limits on this item of expenditure. In my interviews, I noticed a strong concern to keep food expenses to a budget, even if household income had increased substantially over the years (see National Credit Regulator, 2012).[3] Food generally accounts for a smaller proportion of middle-class budgets than transport, for example; but its cost may be varied more flexibly. Some of the less well-off members of my sample complained about the rise of food prices over the last few years, claiming that food is their main expense, absorbing up to 40% of their income in some cases. Regardless of its budget share, most people I interviewed were keenly aware of food prices and could calculate their family's food costs – 'I spend on average R3500 to 4000 per month for the whole family' (Lameez, Indian, married, housewife, mid-thirties, two children). South African women's magazines read by my informants have large sections offering advice to households, especially on food – not only how to cook, but also how to spend your money wisely on food. *You*, for example, receives praise for its cooking section, where the price of a meal is itemised or a weekly food budget is suggested. The monthly magazines *Bona* and *Your family* have a middle-class readership, the first mostly black women and the second women of several races; sometimes they also provide budgets for meals and recipes.

Everyone takes advantage of the cost reductions made possible by large-scale marketing outlets, relying on end-of-month 'specials', bulk-buying and comparing prices. They go for wholesalers like Makro or 'hampers' in Fruit&Veg and buy in bulk at the end of the month, creating a noticeable 'month end payday buying frenzy' in the shopping centres. Usually the people I met buy some food for the whole month – basics such as maize meal, oil, onions, rice and some meat – the weekend after they receive their salaries (around 25th or 27th of the month); while small purchases of bread or milk are made daily or weekly. They are ready to drive far for a good bargain after careful and frequent screening of the 'specials' in mailboxes and newspapers (sometimes distributed to the kids at school): 'I check out all the promotional literature by the supermarkets to identify specials; I plan my purchases accordingly and then I stock up on them!' (Bronwyn, coloured, married, early thirties, one child). She has a system of envelopes with money linked to specific goods, what Zelizer (1997) calls 'earmarking'; for Bronwyn, money is not just an abstraction. A few months after this interview, she and her husband bought a small house in Berea and she left her job as a clerk at a police station to take care of their little daughter full-time. Some of the

---

[2]There is a huge literature on food and cooking. I will mention here only the sources I used directly in analysing the results presented.

[3]No-one takes out a loan to finance food purchases, or at least will confess to have done so. This contrasts with the article 'South African consumers borrowing for basic needs' published in the *Mail&Guardian*, 24 October 2012. http://mg.co.za/author/penwell-dlamini Accessed 12 March 2014.

women participate in *stokvels*, or more specifically in saving clubs that aim to buy groceries in bulk at the end of the year, in order to be able to prepare for Christmas celebrations without too much strain. These societies give an empowering feeling of being in control of your expenditure, while generating solidarity between women who share the plight of having to buy food for their families on a tight budget (Counihan, 1998).

Meat receives special treatment. This accords with what anthropologists know about human relations with animals and not least with their death. Meat is particularly celebrated in South Africa, especially by the whites when grilled on a barbecue. Most meat is bought in a butcher's shop, often in bulk like half a lamb at Bluff Meat, for example. Many black consumers do not hesitate to return to their former townships at the weekend to buy meat or to cross the city for the sake of a special shop. The same is true of my informants in France, who choose carefully where they buy meat. My informants argue that each community has its preferred meat and cooking style and they sometimes reproduce stereotypes, claiming that blacks like to eat cow tripe with dumplings cooked in various ways; the whites go for chicken *à la king* or pork roast and sausages; and chicken and lamb curry are the Indians' favourite. But when faced with the rising cost of food, many of them were inclined to cut back on buying meat.

Despite being a port and seaside resort, Durban has very few fish shops, since its consumption has always been quite low. This paradox is related to the food traditions of the different local groups: Indians, whites and coloureds traditionally consumed more fish than the Zulu majority. Fish and seafood are as a result mostly imported from Mozambique, often sold frozen and quite expensively. Even so, probably as part of a movement towards healthy eating, everyone now says that they eat or would like to eat fish. But most of my informants claimed to know little about how to cook it and preferred to eat fish in restaurants or as a takeaway. Ocean Basket is a favourite among all groups, where the family can eat fish and shrimp after shopping in a mall.

Young couples usually go shopping with their partner, but Maureen (black, divorced, late forties, with five people to feed) goes with her older daughter; other women with their sister or mother. Often also people shop for other members of their family who are not living with them, especially parents who still live in the townships or elsewhere. These monthly food shopping expeditions often end up with a lunch in one of the numerous fast-food chains or restaurants, which are located in shopping centres or malls, as a pay-off for undergoing the duty of shopping. Nevertheless, Durban people do not very often go out to restaurants, either because it is expensive or because they prefer to stay home at night, possibly for security reasons. One restaurateur told me that no more than 40 000 of Durban's three million inhabitants ever go out to eat in the evening, to what he called a 'proper restaurant' – with a chef and not part of a chain. They are more likely to go out for lunch because they can take the children.

2.2 Take away and eating out: pushing the boundaries?

If meat is each community's 'soul food', the women I interviewed described the same range of meals as their favourites for home cooking: macaroni cheese, roast chicken, 'lamb pie', lasagne and 'cottage pie'. Black women add 'pap & meat' to this list. Apart from the recent rise of Italian food, a lot of these recipes share a British colonial heritage (Leong-Salobir, 2011). Bronwyn, who claims to be coloured, says that as a result of being 'in between' she has mastered all the community specialities from *putu pap* with meat and *chakalaka* to Indian curry.

Although most South African women's magazines address a specific community readership, their cooking recipes are remarkably similar, varying mainly in the cost of recipes aimed at readers of different class, despite the fact that in other respects (fashion, advice columns, etc.) their orientation is directed to one group only. Except for one element, the fat used in recipes: cooking recipes in African women's magazines never specify butter, always margarine. French historians such as Braudel (1961) and Febvre (1961) have noted the distinctive use of fats there, historically and geographically; and Mauss (1926 [1947]) has a general discussion about the use of butter (fresh and rancid) in the Mediterranean area. There are in fact two Frances: the France using butter in the north, and the France using olive oil in the south! Are there two South Africas?

My informants are also open to trying new kinds of food, making experiments, taking advantage of industrial food options and chain restaurants (including take away). But at the same time they stick quite closely to the food habits of their own communities; and some people are quite suspicious of what they consider to be 'European' or 'western' food, such as pizza and sushi. Nevertheless, cooking styles are diffused by restaurants, above all as take-away dishes that people are very keen to try. At lunch time, queues line up in front of take-away stands that often sell the same dishes: chicken or mutton curry, grilled chicken, macaroni bolognaise, rice with minced meat, sausages and sometimes fish and chips. Once a week, at least, the people I met and sometimes observed ate take away as a family meal on Friday or Saturday evening, bought at Wimpy, Steers or KFC or at local Indian or Italian take-away restaurants: 'We would go to KFC because my wife is paid on Friday' (Thutukani, black, married, early thirties) or 'I go to a mall where there are different take away restaurants, to buy different food, a pizza for my husband and son, a curry for me ...' (Heidi, white, married to an Indian, mid-forties, three children). Few people go to a 'proper' restaurant as a couple or with a group of friends; sometimes they take the opportunity of offers, and thus some Durban restaurants offer half price for women only. People also often discover restaurants through corporate events. Another way to eat out and to try new recipes is through church and community meals, very often based on a 'bring and share' principle.

The range of choice in South Africa, including Durban, is wide, extending to international chains like KFC, which is very popular, and Wimpy, national chains like Steers (founded by a Greek South African and now everywhere) or Nando's (a chicken peri-peri specialist founded by Mozambicans and common in Britain and Asia); or again there are many local restaurants, mainly Indian, Italian and Portuguese. This range allows South Africans to taste 'globalised' food such as hamburgers or chicken and chips (KFC, Wimpy and Steers), but also more specific recipes, even if these are adapted to local palates: Indian-South African food, Italian-South African food, but also African-South African (Nando's with Mozambican recipes) or a combination of all these influences (Moyo).

In Durban we should add curry, which is thought to be emblematic of the Indian community, even if I was told by some Indian women that they had to learn how to cook it from their in-laws, since cooking in their family consisted mostly of British stews. Curry is nevertheless becoming a regional dish for Kwazulu-Natal because of the large number of take-away restaurants that provide it. In fact, ready-made curry sauces available at the supermarkets are a great success with non-Indian customers who never learned how to mix the spices. Now, thanks to the sachets available, women like Jube and Mpumie (black, sisters, both divorced, both in their late thirties, one with two children) can now prepare their weekly curry. In this way, one community's speciality has become a staple of South African cuisine for all middle classes in Durban.

This Durban lower middle class is pushing out some of the boundaries of food preparation and consumption beyond their former silos. This process is made easier because, at this level of South African society, cooking is not configured by a strong contrast between high and low cuisine (Goody, 1982). I would hazard the generalisation that, in South Africa, there is no cooking style that is reserved to one class or community but not the others. It just is not the case that only members of a particular cultural tradition (or professionals, usually men) know how to cook its food. I return to the question of South African national cuisine in my conclusions.

3. Shopping malls aas shared social spaces

Food is bought mostly in supermarkets, which are usually located in shopping centres or malls. During the late apartheid era, access to mass consumption in the form of shopping malls was largely restricted to whites. These malls – along with the development of tourism – represented one of the few investment channels open to white capital within the country, given the restrictions on its export imposed by the boycott. Since the ANC came to power, shopping malls have sprung up everywhere, reaching the whole population. Indeed President Zuma has been quoted as holding a vision of economic democracy that would grant every village its own shopping mall! There does not seem to be much public discussion in South Africa about how malls erode the base for local shops. At the same time, the government has made a big effort to improve service provision in the townships themselves. In collaboration with the private sector, they have launched shopping malls of a standard and scale equivalent to those built in the most affluent countries. In the Durban area, the relatively modest Umlazi Mega City comes to mind or the more ambitious Bridge City Shopping Centre in KwaMashu. In Chatsworth, an Indian township, the introduction of a new shopping mall provoked a raging controversy over ethnic ownership (Vahed & Desai, 2013). In any case, access to mass consumption – regardless of racial category – has become a central plank of economic and social policy in South Africa. And anthropologists, following Miller's (1998, 2001) example, have shown that middle-class people often judge the quality of life in their neighbourhoods by the shops. I remember Lindiwe telling me: 'It was wonderful when I first arrived here. Before it could take me 45 minutes to find a decent shop, but here they have everything! I never realised what an easy life the people here enjoy' (black, married, early forties, three children). She moved with her family to New Germany from Umlazi Township in 2003.

Malls, shopping centres and supermarkets, which are dominated by a few national chains, represent modern values, including hygienic standards. These mixed commercial spaces are leisure zones comparable in Durban with the seafront promenade. The great suburban malls there like Gateway and Pavillion are places for a family excursion at weekends, for stocking up on food before sharing a meal in a chain restaurant. The mall crowds are drawn from all South Africa's 'communities'

(Houssay-Holzschuch & Teppo 2009). But the clientele of Workshop, for example, in the city centre near a transport hub for taxis, is overwhelmingly black and young, whereas the Musgrave centre (Durban's oldest, in a formerly white area) now attracts different groups at different times of the day or week: a white middle-class crowd on working days, teenagers of all colours after school and more African and Indian families at the weekend. Durban city centre has its fair share of open markets and street stalls, often in areas associated historically with one community; but the people I interviewed went to these places only occasionally.

Middle-class consumption takes place in malls whose character is distinctively national and there is no question that being a citizen of the new South Africa means participating in the forms of sociality and anonymity afforded by shopping centres even more than actually buying stuff there. They are synonymous for many with modernity and abundance. In any case, these malls with their supermarkets frame middle-class consumption as national South African practice. A comparative perspective might allow us to see how participation in mass consumption by South Africans other than whites, after a long period when the country was closed in on itself, may be a way of affirming one's political stance there, of declaring an attachment to the nation while at the same time making links with the rest of the world through buying international brands. In a study of Jewish immigration to New York from Eastern Europe at the beginning of the twentieth century, Heinze (1990) showed how participation in the early stages of American mass consumption was one way in which this community expressed its attachment to the new society. The abundance of commodities there was – and probably still is – a marker of North American identity. South Africa's malls, where many people go just to hang out without necessarily buying anything, show off an abundant world of consumption, within arm's reach.

If everyone likes the shopping facilities of malls, we may wonder whether they can give people the social life they want or a sense of identity comparable with being located in a specific neighbourhood. Many of my black and Indian informants go once a week to shop for specific items in their old neighbourhood. Bronwyn, who lives in formerly white Berea but was born in Wentworth, a coloured neighbourhood, goes back there to shop for meat, to have her hair done and to go to church. The newspapers often stigmatise this kind of behaviour, describing a tension between middle-class aspirations that they consider to be westernised and this attachment to one's 'cultural roots' (Chevalier, 2010; for a French example, see also Chevalier, 2007). Simon Wood's (2011) documentary *Forerunners*, shown at the Durban film festival in July 2011, addresses this question through stories taken from South Africa's emerging black middle class. The director asks people in front of the camera to compare 'the traditional views of their childhood with western consumerism that rules their professional lives' (according to his introduction). Modernity is clearly represented as the world of the city and contemporary consumption. The film reveals in a nuanced way the explorations of four individuals. This kind of reflection on the meaning of tradition and modernity is important for South African research if we wish to place the phenomenon of the new middle class in a comparative perspective. The modern/traditional contrast, not least in this film, hinges always on an opposition between the city, the mall and upmarket consumption – a world associated with white colonialism – and a rural African world. Such a vision recapitulates the myth that South Africa's cities were invented exclusively by whites, thereby implicitly endorsing a claim that behaviour in the country's new social spaces should conform to white standards (Ballard, 2004).

This last example leads me again to ask how people define belonging to this lower middle class: some of my black and Indian informants were adamant that it was the chance to have a house and food, then to have access to other goods and services, often shared with your family. Priya (Indian, married, late thirties, living with her divorced sister and her son) was very specific:

> We are middle class because we are above the bread-line, we can afford food, a flat and two cars. Sometimes we can afford e bit more, a piece of cloth for me, a toy for my nephew who lives with us. But we can't afford to go to the restaurant everyday like the upper-classes. We can vary our diet but we need to budget at the same time, we can't buy everything we want.

To be able to afford a little more than the basics makes the difference, just as to have some consumer purchasing power is associated with agency. But this feeling of upward social mobility is often linked to fear of losing it, which probably distinguishes this from other social classes. The people I interviewed still cling more to their community of origin for self-identification than to any middle-class identity, which is in any case fragile. They know that their power of consumption is indeed quite limited even if it is used to define them as a class.

My black informants also often felt that they were novice consumers, because shopping and access to services more generally were so limited before. This situation recalls historical accounts of the birth of mass consumption in the United States and Europe in the late nineteenth and early twentieth centuries, when the middle and working classes had to learn how to consume (Strasser et al., 1998; Zelizer, 1997). This literature also emphasises the role of advertising and brands in the formation of consumer behaviour. The people I interviewed, especially those aged over 40, expressed an attachment to brands that had been around for a long time: 'because we have always seen them around us'. Knowledge of these brands was a kind of shopping guide for many consumers. Global brands build links to the outside world. Also, the principal victims of segregation have little capital, savings or inheritance to draw on and must go into debt if they wish to consume on any scale (James, 2012). As one of them told me: 'The apartheid system made sure that we knew little about money, since we were supposed not to have any!' (Muntu, black, widowed, late thirties, one child).

The point of comparing South Africa today with historical examples elsewhere is not to imply a unilinear process of modernisation, but to identify questions that might not otherwise be raised. Thus, who is teaching South Africans how to consume and through which media? When I analyse the cooking recipes and advice in women's magazines, I am struck by the sheer prominence there of big companies and retail chains.

4. Consumer behaviour: a political question?

South African marketing professionals have lately been looking for a device that would allow them to describe consumers without using the system of classification prevalent under apartheid. The idea of a Living Standards Measure was perfect for this purpose since it allowed them to define social categories without mentioning a person's race (Burgess, 2002). In consequence – which is just the first paradox of this conflicted practice – it has been widely adopted as a means of identifying the lifestyle of a social group whose only common characteristic is the colour of their skin. When it became impossible to abstract completely from race, which after all had been synonymous with class as a marker of individual identity for over a century, it was replaced by the term 'culture', conceived of as a mark of collective belonging. Also, the Living Standards Measure makes no reference to locating consumers in space: interviews with professionals in charge of marketing automobiles, for example, revealed that this new middle class was taken to be nationally homogeneous and no interest was shown in any regional differences that might exist. The second paradox, however, is that the same marketing professionals invented another word for this class, the black diamonds, which reintroduced race by the back door and was then taken up avidly by the media. The first publication from the Unilever Institute to use this term was its '2007 survey: On the move' (University of Cape Town's Unilever Institute of Strategic Marketing and TNS Reseach Surveys, 24 May 2007). This study was based on interviews with 750 adults in South Africa's main cities drawn from categories 9 and 10 of the Living Standards Measure (Krige 2011:297f). When it comes to projecting images of the new consumer classes, the media are much more influential than social scientists (Chevalier, 2010). Indeed, the latter often draw on labels created by professional specialists for their own analyses. No representation of this sort has been more debated in the South African media than the black diamonds. The term has also found its way into literature; for example, as the title of a work by the writer Zakes Mda (2009) who offers in his novel a nuanced and complex vision of the new urban black middle class without any element of moral condemnation. Since May 2011 it can even be found as a heading in Wikipedia, which describes it as both a 'racial term' and 'pejorative'.

Walter Benjamin (2006), when writing about the construction of urban social types in Europe, refers to moral portraits typically produced by a 'panoramic literature' as a 'phantasmagoria'. He applied this metaphor, taken from a kind of magic lantern show, to Paris's shopping arcades, built at the birth of modern capitalism, which he saw as spectacular theatres of the new commodity culture. In contemporary South Africa, there is no shortage of shopping malls or of moral portraits depicting their denizens. The lifestyle categories that circulate as a result make up for their lack of scientific status by the sheer volume of their public iteration. The stereotype of the black diamonds, built up around an imagined lifestyle, goes with a matching set of 'moral' characteristics that are reminiscent of the phantasmagoria.

Moral portraits of the black diamonds rely mainly on descriptions of consumer behaviour. The long history of economic inequality has yielded a model of cultural legitimacy in which 'good practice', with reference to money and consumption in particular, is implicitly taken to be white. An illusion that diverts and distracts us from reality,[4] the black diamond label masks what is really at stake in society and politics; in this case, the persistence of racial stigma and of a behavioural standard based on the white minority. Morally loaded descriptions range between two poles: anxiously positive at one end; disapproving, even stigmatising at the other. The first extreme celebrates the new middle class, while passing over the conditions of its emergence,

---

[4] Benjamin's critique of capitalism as a dream, as famously developed in the *Passagenwerk*, known in English as *The Arcades Project* (Benjamin 1999), lies beyond the scope of this article.

especially the racial dimension contained in the label itself. The second depicts an alienated group forgetful of its social obligations after discovering the joys of money and consumption. To borrow the terminology of two sociologists close to Bourdieu, media discourse is sometimes populist, sometimes 'miserabilist', more often the latter (Passeron & Grignon, 1989).

So what are the black diamonds supposed to get up to? Their lifestyle apparently leaves little to be desired when compared with the 'conspicuous consumption' of the *nouveaux riches* Americans described by Thorstein Veblen (1992/1899); for these consumers, prestige was the only consideration.[5] In short, their consumption is conspicuous and is described as such, not only by the media but also by the specialists I interviewed (marketing and advertising experts). Some articles express reservations about it all: ostentatious display of wealth by the new black middle class is thought to be in bad taste for a country with so many poor people. Against the moralisers, blogger Sentletse Diakanyo[6] has come to the defence of conspicuous consumption on his fellow-citizens' part by appealing to the central idea of Adam Smith's *Wealth of Nations* (1961 [1776]), namely that the pursuit of individual self-interest is the motor of economic development. At the same time, he draws on a psychological argument produced by a study of black and Hispanic consumer behaviour in the United States: conspicuous consumption is a response to historical deprivation by groups who wish to affirm their new social standing. Comments on this blog post were mixed; but their sheer number shows the level of interest among his readers, most of whom had little time for the moralisers. Even though my informants are lower middle class, the ubiquity of the term black diamonds and the descriptions attached to it affect them too, since the distinction between lower and upper middle classes is usually collapsed in such discourses. I should, however, point out that my informants often rejected application of the term 'middle class' to themselves, preferring to be known as a 'professional' class. They are teachers, nurses, technicians, librarians, secretaries, police officers, independent consultants, and so forth. There was such an African professional class long before 1994, of course. It is not an invention of 'democracy'.

5. Conclusions

I am less concerned here with the construction of an urban social type or with defining class according to income – which is in any case hard to find out – than with trying to understand how the middle class might be identified through common practices, in the home as much as in newly shared social spaces.

If Durban's middle classes do not define themselves through their eating and cooking habits, these practices have become quite similar across the boundaries between communities. The fast-food chains facilitate this process of self-identification at a number of levels; through the food styles on offer, they make South Africa at once an international and a national, even African, place. Restaurants allow one to eat a neighbour's food without knowing how to make it oneself – curry or fish, for example – and I would add without necessarily meeting their neighbours; which is a general

---

[5]Conspicuous consumption has its own television programme in South Africa, *Top Billing* (SABC3), every Thursday evening, plus a magazine of the same name. This programme shows luxurious domestic interiors and celebrations (above all marriages) with protagonists from all of the country's communities.
[6]http://www.thoughtleader.co.za/sentletsediakanyo/2010/12/03/in-defence-of-conspicuous-consumption Accessed 17 April 2014.

paradox inherent in consuming commodities. The range of possibilities on offer through manufactured food and restaurants no doubt contributes to the development of a national taste, but also to a regional taste specific in this instance to Kwazulu Natal. I would not go so far, however, as to say that a specifically South African cuisine has yet emerged from all these culinary encounters, certainly not anything that we could call a creole food culture (Wilk, 2006a, 2006b).[7] I have also been struck by how the experts, magazines, television shows, books and food fairs advance recipes that both take into account South African 'reality' and help to build a supra-regional universe consisting of English speakers living in the southern hemisphere. This cuisine is built on recipes using a similar range of ingredients and promoted by chefs from Australia and South Africa in particular. Following Appadurai (1988), one might say that they are building a new ethnoscape on the basis of shared food tastes. At the same time each group has its 'soul food', usually focused on a highly-charged symbolic food – meat – although the penchant of white settlers for outdoor barbecues is a common element for some.

Analysis of consumption patterns, taking into account historical context and generational difference, and especially of food consumption, because it is literally incorporated so deeply into social life, helps us to understand better the articulation and transformations of race and class identities that for so long were inseparable. These transformations contribute also to a new definition of regional and South African identity. Consumption is rightly considered to be a political question. It should not be left as the exclusive province of lifestyle professionals, such as marketers and journalists.

References

Appadurai, A, 1988. How to make a national cuisine: Cookbooks in contemporary India. Comparative Studies in Society and History 30(1), 3–24.

Ballard, R, 2004. When in Rome: Claiming the right to define neighbourhood character in South Africa's suburbs. Transformation 57, 64–87.

Benjamin, W, 1999. The Arcades Project. Harvard University Press, Cambridge, MA.

Benjamin, W, 2006. The writer on modern life: Essays on Charles Baudelaire. Harvard University Press, Cambridge, MA.

Braudel, F, 1961. Alimentation et catégories de l'histoire. Annales. Économies, Sociétés, Civilisations 16e(4), 723–8.

Burgess, SM, 2002. SA tribes: Who we are, how we live and what we want from life in the new South Africa. David Philip, Cape Town.

Chevalier, S, 1997. L'idéologie culinaire en Angleterre ou comment séparer le blanc du jaune in Pratiques alimentaires et identités culturelles. Ethnologie Française 1(XXVII), 73–9.

Chevalier, S, 2007. Faire ses courses en voisin: pratiques d'approvisionnement et sociabilité dans l'espace de trois quartiers de centre ville (Paris, Lyon et Besançon). http://metropoles.revues.org/107 Accessed 25 September 2014.

Chevalier, S, 2010. Les Black Diamonds existent-ils? Médias, consommation et classe moyenne noire en Afrique du Sud. Sociologies Pratiques 20, 75–86.

Chevalier, S, 2012. Comment partager les mêmes espaces? Les classes moyennes à Durban (Afrique du Sud). Espaces et Sociétés 148–9(1), 129–44.

Chevalier, S & Escusa, E, 2011. Les pratiques alimentaires des classes moyennes sud-africaines, Rapport de recherche pour l'entreprise Bel, Septembre, Entreprise Bel, Paris. 78 pp.

---

[7] I should mention here the cooking book *Indian Delights* (published in 1961 with many revisions and more than 150 000 copies sold, well beyond the Indian community). Its recipes claim to be both typically Indian and different from those of India and Pakistan. This book was written by Zuleikha May for the Women's Cultural Group. Some elements of African cooking, such as a 'curry-pap', are integrated into the recipes. See also Vahed & Waetjen (2010) and Wardrop (2012).

Counihan, CM, 1998. Introduction: food and gender: Identity and power. In Counihan, CM & Kaplan, SL (Eds.), Food and Gender: Identity and Power. Harwood Academic, Amsterdam, pp. 1–11.

Febvre, L, 1961. Essai de carte des graisses de cuisine en France. Annales. Économies, Sociétés, Civilisations 16e(4), 747–56.

Goody, J, 1982. Cooking, Cuisine and Class: A Study in Comparative Sociology. CUP, Cambridge.

Heinze, A, 1990. Jews Immigrants, Mass Consumption and the Search for American Identity. University of Columbia Press, New York.

Houssay-Holzschuch M & Teppo, A, 2009. A mall for all? Race and public space in post-apartheid Cape Town. Cultural Geographies 3, 351–79.

James, D, 2012. Money-go-round: Personal economies of wealth, aspiration and indebtedness. Africa 82(1), 20–40.

Krige, D, 2011. Power, identity and agency at work in the popular economies of Soweto and Black Johannesburg. Unpublished PhD dissertation, Social Anthropology, University of Witwatersrand, Johannesburg.

Leong-Salobir, C, 2011. Food culture in colonial Asia: A taste of empire. Routledge, London.

Mauss, M, 1926 [1947]. Manuel d'ethnographie. Payot, Paris.

Mda, Z, 2009. Black diamond. Penguin, London.

Miller, D, 1998. A theory of shopping. Polity Press, Cambridge.

Miller, D, 2001. Dialectics of shopping. University of Chicago Press, Chicago.

National Credit Regulator, 2012. Consumer credit market report, First Quarter March 2012. http://www.nca.org.za Accessed 29 September 2014.

Passeron, J-Cl, and Grignon, Cl, 1989. Le savant et le populaire. Misérabilisme et populisme en sociologie et en littérature. Seuil, Paris.

Smith, A, 1961 [1776]. An Inquiry into the Nature and Causes of the Wealth of Nations. Methuen, London.

Strasser, S, McGovern, C & Judt, M, 1998. Getting and spending. CUP, Cambridge.

Vahed, G & Desai, A, 2013. Trading in the township: A checkered history. In Desai, A & Vahed, G, Chatsworth, the Making of a South African Township. UKZN Press, Durban, pp. 349–60.

Vahed, G & Waetjen, T, 2010. Gender, Modernity and Indian Delights: The Women's Cultural Group in Durban, 1954–2010. HSRC, Cape Town.

Veblen, T, 1992/1899. The theory of the leisure class. Transaction publishers, New Brunswick.

Wardrop, J, 2012. Speaking out loud: Muslim women, Indian Delights and culinary practices in eThekwini/Durban. Social Dynamics 38(2), 221–36.

Wilk, R, 2006a. Home Cooking in the Global Village: Caribbean Food from Buccaneers to Ecotourists. Berg Publishers, Oxford.

Wilk, R, 2006b. Fast food/slow food: The cultural economy of the global food system. Altamira Press, Lanham, MD.

Wood, S, 2011. Forerunners. Film, South Africa, 52.

Zelizer, V, 1997. The Social Meaning of Money. Princeton University, Princeton, NJ.

Index

Note: Page numbers in *italics* represent tables
Page numbers in **bold** represent figures
Page numbers followed by 'n' refer to notes

Printed and bound by CPI Group (UK) Ltd, Croydon, CR0 4YY

21/10/2024

01777095-0011